Evolutionary physiological ecology

Evolutionary physiological ecology

Edited by
P. Calow
Department of Zoology
University of Sheffield

Cambridge University Press
Cambridge
New York New Rochelle
Melbourne Sydney

CAMBRIDGE UNIVERSITY PRESS
Cambridge, New York, Melbourne, Madrid, Cape Town, Singapore, São Paulo, Delhi

Cambridge University Press
The Edinburgh Building, Cambridge CB2 8RU, UK

Published in the United States of America by Cambridge University Press, New York

www.cambridge.org
Information on this title: www.cambridge.org/9780521320580

First published 1987
This digitally printed version 2008

A catalogue record for this publication is available from the British Library

ISBN 978-0-521-32058-0 hardback
ISBN 978-0-521-10165-3 paperback

CONTENTS

CONTRIBUTORS

(First authors only)

B.L. BAYNE
Institute for Marine Environmental Research, Prospect Place, Plymouth, PL1 3DH, U.K.

P. CALOW
Department of Zoology, University of Sheffield, Western Bank, Sheffield S10 2TN, U.K.

E. GNAIGER
Institute Zoophysiologie, Universitat Innsbruck, Peter-Mayr-Strasse 1a, A-6020, Innsbruck, Austria.

J.P. GRIME
Unit of Comparative Ecology (NERC), University of Sheffield, Western Bank, Sheffield S10 2TN, U.K.

T.B.L. KIRKWOOD
National Institute for Medical Research, The Ridgeway, Mill Hill, London NW7 1AA, U.K.

A.L. KOCH
Department of Biology, Indiana University, Bloomington, Indiana 47405, U.S.A.

M. LYNCH
Department of Ecology, Ethology and Evolution, Shelford Vivarium, 606 E. Healey Street, University of Illinois, Champaign, Illinois 61820, U.S.A.

B.K. McNAB
Department of Zoology, University of Florida, Gainesville, Florida 32611, U.S.A.

R.M. SIBLY
Department of Zoology, University of Reading, Whiteknights, Reading, RG6 2AJ, U.K.

L.B. SLOBODKIN
Department of Ecology and Evolution, Division of Biological Sciences, State University of New York at Stony Brook, Stony Brook, New York 11794, U.S.A.

O.T. SOLBRIG
Gray Herbarium, Harvard University, 22 Divinity Avenue, Cambridge, Mass. 02138, U.S.A.

R.LeB. DUNBRACK
Department of Fisheries and Oceans, Pacific Biological Station, Nanaimo, B.C. V9R 5K6, Canada.

PREFACE

This book is based on a symposium that took place as part of the Third International Congress of Systematics and Evolutionary Biology (ICSEB III) at the University of Sussex (Brighton, U.K.) in July 1985. It is therefore certainly a Proceedings, but having said that all the chapters have been refereed by at least 2 people and, as indicated in the first chapter, some attempt has been made to structure the material into a framework that develops a coherent and biologically-reasonable theme. My thanks go to the contributors for their patience and understanding in helping to make this at all possible, but the responsibility for any shortcomings in the organization and presentation of the final version must, of course, rest ultimately with the editor!

Thanks are also due to the Organizing Committee of ICSEB III for supporting the symposium and to the Cambridge University Press for supporting this publication. Special thanks are due to Angela Saxby for often doing better than Maxwell's Demon, creating order out of chaos, in typing up all the material into camera-ready copy. Lorraine Maltby prepared the index and did last-minute proofreading.

PC

(Sheffield, 1986)

EVOLUTIONARY PHYSIOLOGICAL ECOLOGY?

P. Calow

INTRODUCTION

Despite an early, probably anthropocentric view of physiology that "there has been no evolution of function, all living things have certain fundamental metabolic activities" (Woodger, 1929; see also Reid, 1985), both comparative and environmental physiology have subsequently exposed considerable metabolic variation that appears to be adaptive. So "evolutionary physiological ecology" is something of a tautology; the "evolutionary" is redundant. The reason for wanting to emphasise it here, however, is to signal a change in orientation from an approach that has largely been concerned with correlating metabolic with ecological variation on a temporal and spatial scale to one that attempts to formulate adaptive explanations of metabolism more explicitly, often on the basis of rigorous models. This owes much to the influence of behavioural ecology where optimality techniques have been applied in the construction of foraging models (Krebs & Davies, 1984).

In what follows the first 5 chapters are concerned with processes and the rest with examples of physiological adaptation at an organismic level. Model making is a pervasive feature of all sections, particularly the first; but an equally important message is that it is only possible to build realistic models and to pose relevant questions on the basis of a considerable amount of physiological and ecological information. A number of the chapters make this point but none more clearly and explicitly than that by Grime et al. (Chapter 7). They advocate a classification of physiological and other traits in plants that attempts to identify suites of traits, broad strategies, that transcend taxonomic affinity.

ECONOMIC ORGANISMS

A useful starting point for evolutionary physiological ecology (that has its roots in ecological energetics; Phillipson, 1966) is to view the organism as a device that allocates resources amongst a number of metabolic demands - basically, those that generate power for physical and biochemical work and those that synthesise biomass. The amount of resource available for allocation might vary from one organism to another, but only within limits, so we can assume that the resources available for allocation are limited and this means that if more resource is used in one aspect of metabolism less is available for others - this is the so-called Principle of Allocation first hinted at by Cody (1966). Clearly, allocation patterns will profoundly influence the physiology of organisms (and also their behaviour and even ontogeny; Townsend & Calow, 1981, Calow 1984) but it is also obvious that these physiological consequences will, in turn, influence all the key components of fitness - survivorship, fecundity and developmental rate. Each of these should, according to NeoDarwinian principles, be maximized but the Allocation Principle suggests that this could not occur in all components simultaneously. Hence, it becomes appropriate to consider solutions that compromise between the physiological trade-offs. The optimum compromise depends upon the form of the trade-off (because this fixes the relationship between fitness components) and the demographic circumstances in which it has to function (because these will determine whether survivorship, fecundity or developmental rate are likely to have most impact on overall fitness). Trade-offs can be approached from a physiological perspective (e.g. Calow, 1985) and demography from an ecological prospective; so physiology is not only enriched by an ecological/evolutionary analysis, but it can itself also make important contributions to the evolutionary analysis.

Another possible trade-off, based more on thermodynamic constraints than the Principle of Allocation, is between metabolic rate and efficiency. This has been hinted at, loosely, in a number of previous publications (e.g. Calow, 1984) but is addressed, here, rigorously by Gnaiger in Chapter 2 using modern thermodynamic theory.

DIRECT AND INDIRECT MEASURES OF FITNESS

Ideally, in framing models and applying optimization procedures it is necessary to map physiological processes on to their demographic consequences. This mapping will vary with the age of organisms so that it is possible to represent fitness multidimensionally, in terms of age-specific, physiologically-dependent fitness components and to imagine selection operating as though searching through this space for combinations that maximize fitness within the limitations set by the trade-offs discussed above. But this is far too complex to handle in practice. We therefore often have to decompose fitness components into juvenile and adult elements and consider their effects on fitness in pairwise combinations. This leads to adaptive or selective landscapes (Sibly & Calow 1983) and involves defining: (1) the relative contribution each component makes to fitness, so that the form of the landscape can be specified and combinations giving the same fitness, isoclines, can be mapped on to them; (2) which combinations of fitness components are biologically/physiologically feasible so that a trade-off curve can also be mapped on to the landscape; (3) how either the isoclines or the trade-offs are influenced by the ecology of the animal; (4) the optimum physiological investment for particular ecological circumstances by reference to the elements in (1) to (3).

This, of course, is the ideal, and is advocated in Chapters 3 & 4. Rarely, however, can it be achieved in practice. A common problem is the mapping of physiologies into their demographic effects because physiologies operate on a short-term, second-by-second basis, whereas demographic effects can operate over longer periods of days, weeks or years. Hence more immediate, less direct measures of fitness are used in optimization models. For example, it is often assumed that maximizing production will maximize fitness because this increases developmental rate, increases size for a given age, and ultimately maximizes fecundity. The advantage of models based on this so-called Maximization Principle (Calow 1984) is that they can be framed entirely in physiological terms. The disadvantage is that they often ignore survivorship consequences of metabolic strategies. They have nevertheless been used widely in the literature, for example in the development of optimal foraging theory (Krebs & Davies, 1984), and are used here in Chapter 2 and throughout Chapters 6 to 12. In Chapter 11, Dunbrack and Ware combine a model based on the Maximization Principle with one that incorporates survivorship considerations.

GENETIC BASIS

The assumption in all the foregoing is that physiological processes and the trade-offs between them have a genetic basis. This is a reasonable assumption in that resource allocation is enzyme modulated and must ultimately be gene specified. Those genes that specify allocation patterns that bring together combinations of fitness components that lead to maximum fitness will, by definition, spread at the expense of the others.

Clearly the precise relationship between genes and their physiological effects and the genetic basis of interaction between physiological effects, trade-offs, is crucial to an understanding of the evolution of physiology. Formally this involves a mapping of the phenotypic adaptive landscape discussed in the last section into the genotypic landscape of Sewall Wright (1932).

This part of the program has barely been started. On the one hand information is needed, of a molecular/genetical kind, on the causal chain from a physiological character to the enzyme controls (as in Hochachka & Somero 1984) and then to genetic specifications (Watt 1985) and this route is followed by Bayne in Chapter 10. On the other hand, information is also needed of a classical quantitative genetical kind on the genetic correlation between physiological traits (Falconer, 1981).

Another important issue is variance in gene-controlled, metabolic traits. Under what circumstances is intra-individual variance (metabolic adaptability or plasticity) favoured over inter-individual variance in biochemical and physiological traits? This is the question raised by Lynch and Gabriel in Chapter 5 and touches on the whole problem of reconciling the considerable genetic variance that has been exposed by electrophoretic techniques with neoDarwinian principles.

IMPORTANCE OF THE ORGANISM

The concept of trade-off reminds us that one gene-determined physiological process has to work in association with others and is subject to natural selection within this organismic context. An allele may express itself differently and have different fitnesses in different genetic backgrounds (Lynch & Gabriel, Chapter 5). It is therefore necessary not only to follow the reductionist pathway of attempting to get some insight by decomposing complex wholes into their component processes, but also to remember the need for a holistic appreciation. Chapters 6 to 12 are largely concerned with this organismic orientation. These address issues ranging from the physiological ecology of prokaryotes and speculations on physiological/biochemical problems associated with the origin of biological, replicating systems (Koch, Chapter 6), through the adaptive significance of plant form and function (Chapters 7 & 8), to symbiosis in Hydra (Chapter 9), size and the foraging behaviour of fishes (Chapter 11) and homoiothermy in mammals (Chapter 12).

REFERENCES

Cody, M.L. (1966) A general theory of clutch size. Evolution, 20, 174-184.

Calow, P. (1984) Economics of ontogeny. In: Evolutionary Ecology, Brit. Ecol. Soc. Symp. No. 23, Ed. B. Shorrocks, pp. 81-104.

Calow, P. (1985) Adaptive aspects of energy allocation. In: Fish Energetics New Perspectives, Eds. P. Tytler & P. Calow, pp. 13-31. Croom Helm, London & Sydney.

Falconer, D.S. (1981) Introduction to Quantitative Genetics (2nd edn.). Longman, London & New York.

Hochachka, P.W. & Somero, G.N. (1984) Biochemical Adaptaptation. Princeton University Press, Princeton, New Jersey.

Krebs, J.R. & Davies, N.B. (1984) Behavioural Ecology. An Evolutionary Approach (2nd edn.). Blackwell Scientific, Oxford.

Phillipson, J. (1966) Ecological Energetics. Edward Arnold, London.

Reid, R.G.B. (1985) Evolutionary Theory : The Unfinished Synthesis. Croom Helm, London & Sydney.

Sibly R. & Calow, P. (1983) An integrated approach to life-cycle evolution using selective landscapes. J.Theor.Biol., 102, 527-547.

Townsend, C.R. & Calow, P. (Eds) (1981) Physiological Ecology. An Evolutionary Approach to Resource Use. Blackwell Scientific Publications, Oxford.

Watt, W.B. (1985) Bioenergetics and evolutionary genetics : opportunities for new synthesis. Am.Nat., 125, 118-143.

Woodger, J.H. (1929) Biological Principles. Routledge & Keegan Paul, London.

Wright, S. (1932) The roles of mutation, inbreeding, crossbreeding and selection in evolution. Proc. VI Int. Congress Genet., 1, 356-366.

OPTIMUM EFFICIENCIES OF ENERGY TRANSFORMATION IN ANOXIC METABOLISM
THE STRATEGIES OF POWER AND ECONOMY

E. Gnaiger

INTRODUCTION

Anoxic Muscle and Mussel

We are all fascinated by high-power phenomena in nature, be they the flash-like jet-propulsion of an escaping squid, the outburst of energy in a hunting cheetah, or the explosive sprint of a racing athlete. The underlying physiological processes have evolved under the selective pressure of optimizing effective power strategy (P-strategy), just as the potential for rapid growth and reckless resource exploitation is rooted in r-selection of species (MacArthur & Wilson, 1967; Gnaiger, 1983a).

A different kind of attraction stems from the vigilant economy prevailing in the living world. Economy-strategy (E-strategy) favours traits that not only use resources effectively, but efficiently. At the maximum input-output efficiency of energy transformation, however, the rate becomes infinitely slow (Prigogine, 1967). Therefore, maximum power output and maximum efficiency are mutually exclusive and consequently P-and E-strategy propagate divergent traits.

My aim is to explain some of the diversity of energy metabolism in terms of the power-economy concept of optimum action (P/E-concept). I will discuss high-power, anoxic muscular exercise, and high-economy, endurance of invertebrates in anoxic environments as characteristic examples. Anoxic metabolism provides the catabolic power at the extremes of low and high metabolic energy flow in animals. Anoxic metabolism is not only phylogenetically the most primitive process of biological energy transformation (Broda, 1975; Livingstone et al., 1983), it is functionally the most versatile mechanism to power maintenance, growth or locomotion in bacteria (Fenchel & Blackburn, 1979; Stouthamer, 1977), plants (Crawford, 1978), and animals including man (Hochachka & Somero, 1984). The physiological functions addressed here, namely locomotion and maintenance, contribute significantly to survival and dispersal, and hence to evolutionary fitness. Somatic and reproductive growth are, like locomotion, energy expenditures above or competing with maintenance requirements. It is therefore reasonable to assume that the energetics of growth follow the patterns of the power-efficiency trade-off, as captured by the model of muscular energetics and environmental anoxibiosis (Gnaiger, 1983a).

First I will outline the distinctive characteristics of the power-economy concept of optimum action by explaining the properties of a simple (one-compartmental) linear energy converter (Kedem & Caplan, 1965). For a functional interpretation of biochemical energy metabolism, we have to take into account the two-compartmental structure of the catabolic-anabolic machinery of cells. A formal derivation of the theory is given in the Appendix (where equations are labelled A1 etc.).

In putting the P/E-concept to the empirical test, I will draw mainly on experimental data gained on vertebrate muscle (di Prampero, 1981; Kushmerick, 1985) and anoxic mussel (Wijsman, 1976; de Zwaan, 1983). By rationalizing the optimum efficiencies of anoxic ATP-production (Gnaiger, 1983a), improved insight is gained into the general patterns of biochemical and physiological adaptation. Some evidence is available suggesting that the P/E-concept can explain the differential constraints on growth rate and growth efficiency (Westerhof et al., 1983; Koch, 1985). Therefore, the distinction between density-dependent selection for maximum epidemic growth (r-strategy) and maximum sustaining capacity (K-strategy) (Boyce, 1984) can be understood as a special case within the framework of the P/E-concept of optimum action.

OPTIMUM VERSUS MAXIMUM EFFICIENCY

Classical thermodynamics defines the maximum amount of work which can, under defined conditions, be extracted from the conversion of a unit amount of substrate. Complete (100%) extraction of Gibbs energy or maximization of efficiency to 100% implies complete reversibility with infinitely slow net rates, that is equilibrium of a stable system. This maximum cannot be optimum for life. On the contrary, the dynamics of life and work depend on efficiency and irreversibility in proportioned balance. A limited reduction and therefore optimization of efficiency is required. Seen from this perspective, irreversibility is not a mere waste of energy but is necessary for optimum system performance in time. P-strategy maximizes the work per unit time which is power (J s^{-1} = W). Think of a 100-m race: maximum power output is equivalent to the minimum time required for a given amount of work; the inevitable cost of minimizing time is rapid exhaustion, i.e. a high dissipation of energy at a low (optimum) efficiency. E-strategy, however, increases the time of sustained submaximum power output by simultaneously increasing the power output per unit power input (i.e. efficiency). The increase in (optimum) efficiency is limited according to the relative selective values of energy and time.

Mathematically, optimum principles are not clearly distinguished from extremum (minimum or maximum) principles (Hildebrandt & Tromba, 1985). For physiological functions and evolutionary success, however, a strict distinction applies: independent variables (fitness functions) are optimized in order to maximize fitness (Fig. 1). Every modification of a fitness function towards the optimum is an adaptation or acclimation. The assignment of

Fig. 1. Adaptation is optimization of a fitness function (horizontal arrows) to maximize a fitness parameter (sloping arrows) under a defined set of constraints.

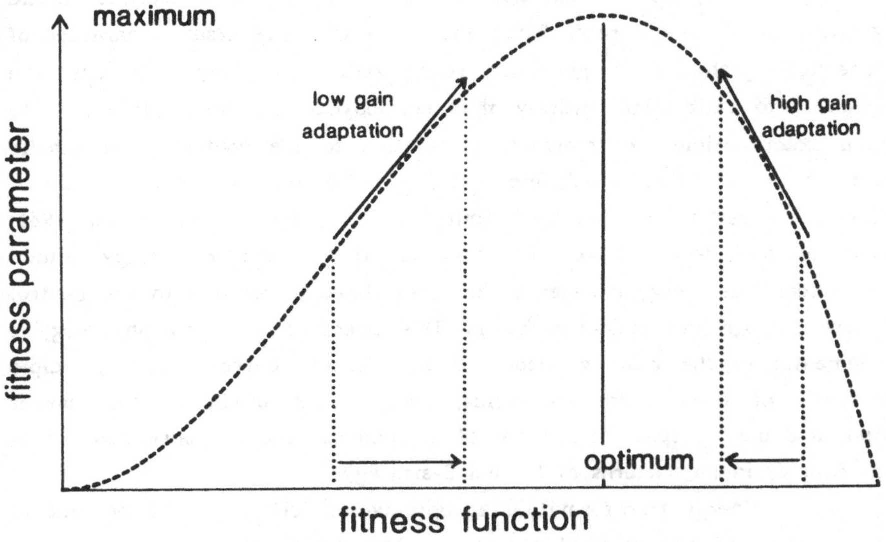

a specific optimum value to the fitness function depends on (1) our hypothesis about the functional significance and resource limitations of the process under study, and (2) on the physicochemical relationship between fitness parameter and fitness function which dictates the need for a "best compromise" (Fig. 1). The distinction between maximization and optimization of a particular variable has important practical and theoretical consequences. However, there is a tendency to consider maximization of a simple and intuitively appealing function as adaptive. The scientific view in an energy-rich culture is likely to be dominated by the paradigm of optimum efficiency for maximum power (Odum & Pinkerton, 1955). Accepting that scientific paradigms follow the trends in society (Kuhn, 1970), we are not surprised that - after the oil crisis - the maximum efficiency of an energy converter (Kedem & Caplan, 1965) has been misleadingly renamed as "optimum efficiency" (Stucki, 1980). Efficiency is a fitness function with an optimum that is different from the maximum in most cases (Fig. 1). By this argument, the calculation of degrees of decoupling which maximize power efficiency at various force efficiencies in mitochondria (Stucki, 1980) is explained as a physiologically irrelevant exercise (for more detailed explanation of terms see below). To retain the physiological significance of optimum efficiency as a fitness function, we must define the environmental conditions and physiological processes which stipulate P- or E-strategy (Gnaiger, 1983a).

Today many biochemists believe exclusively in an enzyme kinetic regulation of metabolic rate. While the necessarily irreversible character of biochemical pathways is generally recognized, they deny explicitly the possibility to rationalize entirely the thermodynamic terms, and hence to define exact optimum efficiencies, in relation to the control of metabolic fluxes (Atkinson, 1977; Newsholme & Start, 1973). However, with the aid of optimization techniques (Maynard Smith, 1978; Calow & Townsend, 1981; Krebs & McCleery, 1984), the concept of P- and E-strategy allows biochemical free-energy changes to be rationalized in relation to the control of flows through biochemical pathways. This aspect of ecological physiology is fundamental to the growing theory of evolutionary energetics. A principle hypothesis of evolutionary energetics, then, is the prediction that power output and input-output efficiencies of biochemical energy conversion follow the strict optimum patterns of P- and E-strategy.

Energy transformation at optimum efficiency is the product of the evolution of energy-coupled systems. The dynamics of such systems is described by an important theory of the "thermodynamics of irreversible processes" (Prigogine, 1967). By definition, conservation of work in the output reaction is a fully reversible process. Functionally, this conservation of energy is the most significant task in cellular energy transformation. Moreover, cells operate as isothermal energy converters, so thermal changes are merely accompanying features in the transformation of chemical energy. In this context, it is logical to replace the terms irreversible and non-equilibrium thermo-dynamics by ergodynamics, defined as the theory of coupled dissipative (irreversible) and conservative (reversible) energy flow. Ergo-(work-) dynamics is that branch of energetics which is concerned with the transformation and performance of work in time, the "motion of energy" (from Greek ergon = work). The coupling of an input force to an output force of equal magnitude but reversed in sign yields 100% efficiency. The net force in such a system is zero: it is in ergodynamic equilibrium. In contrast, physicochemical or thermodynamic equilibrium is defined as a state of minimum energy content. The Gibbs energy of a system in ergodynamic equilibrium is high, since each half reaction of the coupled reaction is maintained away from its partial physicochemical equilibrium by the compensation of input and output forces. A system in ergodynamic equilibrium is fully coupled; the antagonistic input and output forces cannot dissipate via leaks, that is by decoupling. Only fully coupled systems will be discussed here for reasons explained below.

POWER, ECONOMY, AND FITNESS

The power-economy concept (P/E-concept) of optimum action is based on a dynamic interpretation of Gibbs energy changes characteristic for living cells. In the steady state, chemical potential differences between the substrates and products provide the driving force for the metabolic machinery. Without the pressure or drive of a chemical force, the reaction rate is zero even in the presence of enzymes. In a range of low forces and in many enzyme-catalyzed reactions there exists a near-linear relationship between the driving force and metabolic flow (Caplan & Essig, 1983). Therefore a direct relationship exists between chemical force and metabolic power: power is the product of force and flow.

Oxygen, like the fuels carbohydrate, lipid and protein, is a catabolic substrate. Hence, oxygen deprivation implies a state of "starvation" whereby limitation is in terms of chemical energy per mole instead of available amount of organic substrate. The chemical driving force is the molar Gibbs energy of reaction (kJ mol^{-1}) at constant temperature and stable concentrations of substrates and products. Chemical force can be set free either to give rise to chemical, mechanical and other forms of work, or to be dissipated irreversibly as entropy. The proportion of the two antagonistic aspects of the input force, conservation and dissipation, is the efficiency. It characterizes the instantaneous power potential of the system when the kinetic parameters are constant.

The general flow-force relationship applies equally to power input and output of an energy converter, but the input-output net force determines the rate of the coupled reaction. The input force drives the coupled reaction at a maximum rate if it is not compensated by output force, that is at an output force of zero. Then the power input is maximum, yet the power output (output flow x output force) and efficiency are zero (Fig. 2a; f = 0). Low power output despite a high rate of the coupled process is due to inefficient energy conversion. This is functionally advantageous only if the actual fitness parameter is the reaction rate instead of power output (Kedem & Caplan, 1965). The reaction rate, and consequently power input as well as output, are zero at the limit of ergodynamic equilibrium at maximum efficiency (Fig. 2a; f = 1.0). Between these limits, maximum power output is achieved (Fig. 2a). P-strategy involves a compromise between maximum rate and maximum efficiency.

Power strategy produces instantaneously maximum power output of a system. Power-optimum efficiency is 50% in a one-compartmental linear energy converter; that is the input-output efficiency at which power output is maximum, at constant input force (Fig. 2a). P-strategy evolves under circumstances when resource availability is unlimited but resource utilization is constrained by the input mechanism (e.g. aerobic hypoxia), or when instantaneous benefits from power output outweigh the necessary cost of relatively inefficient resource dissipation.

Power input is utilized more economically at higher efficiencies. With increasing output force and force efficiency (Eq. 2.1) the input-output net force is reduced. Efficiency induces an <u>ergodynamic inhibition</u> of the coupled flow. This is why the power input is progressively depressed with increasing efficiency. At high efficiency, the reduction of the rate of the coupled reaction reduces the power output (Fig. 2b; "ergodynamic inhibition"). E-strategy involves a compromise between maximum power and maximum efficiency.

E-strategy optimizes power output at an economical level. The resulting economical power is the product of power output times the value of

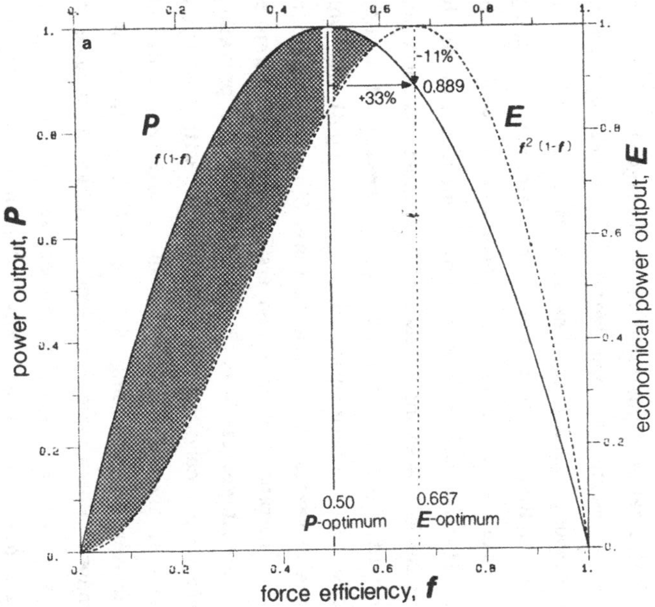

Fig. 2a. Optimization of force efficiency (normalized force ratio, f) to maximize two divergent fitness parameters, power output (P) and economical power output (E = fP), in a one-compartmental, fully-coupled linear energy converter. Plots are normalized relative to the maximum of each fitness parameter at constant input force (Eq. A4.2 and A5). In a fully-coupled system, the force efficiency equals the input-output energy or power efficiency, that is total output divided by total input. Power output is maximum at the P-optimum efficiency, whereas the product of efficiency and power output is maximum at the E-optimum. E-optimum efficiency is increased by 33% while power output is decreased by only 11% in E- relative to P-strategy (arrows).

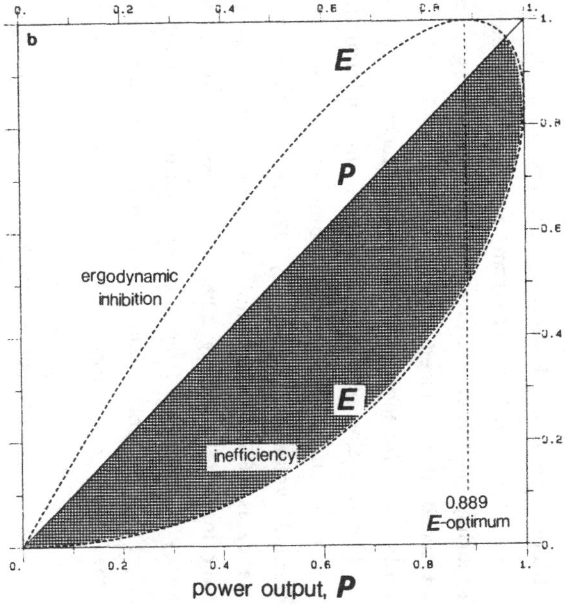

Fig. 2b. Optimization of power output (P) to maximize economical power output (E; projection of E on P from Fig. 2a). At low efficiencies, the low power output and high rate of energy loss yield very low values of economical power (lower broken line, "inefficiency"). E reaches only 50% at a pseudo-optimum power at f=0.33. E is still 16% below maximum due to low efficiency at the maximum power output. At supra-optimum efficiencies, the low power output and low rate of energy loss yield a slow decrease of economical power with decreasing power output and increasing efficiency (upper broken line, "ergodynamic inhibition" of the reaction rate due to a depression of net force).

efficiency (Fig. 2a). In E-strategy, economical power is the fitness parameter that is maximized (Fig. 1) in which case efficiency (Fig. 2a) and power output (Fig. 2b) are optimized fitness functions. E-strategy evolves under limited resource capacity when the long-term benefits from economical resource utilization, and hence enduring potential, outweigh the disadvantages of relatively low power output tuned at a reduced level.

Economical optimum efficiency is a continuous function of the relative selective values of minimizing energy dissipation and minimizing time. A characteristic example of E-strategy is obtained if power is multiplied with the numerical value of efficiency (Stucki, 1980). Then economical optimum efficiency is 66.7% in a simple linear energy converter (Fig. 2a). Whereas the power input falls off by 33%, the power output drops only by 11% as the efficiency is increased from the power-optimum (0.50) to the economy-optimum (0.67; Fig. 2a). This assumes that the input force and the internal conductance (the kinetic properties) of the system are maintained constant.

Mismatched strategies of resource utilization and diminished resource capacity are reasons for the tragic death of starving people. Moreover, it is general ecological and physiological knowledge that rates of growth and locomotion of organisms are frequently constrained by limited environmental or internal resource capacities. Resource capacity differs from instantaneous resource availability, in that the term resource (buffering) capacity incorporates the potential for resource regeneration as well as the potential for recycling or effective waste disposal. Power input draws on the resource capacity. A decrease in power input leads to a proportional decrease in power output at constant input-output efficiency. Increased efficiency, however, yields a decrease of resource dissipation (power input) per unit power output, and therefore increases the endurance time. Since resources are limited, the efficiencies of metabolic energy conversion must be optimized according to the selective advantages of P- or E-strategy under the prevailing environmental and physiological conditions. For systems with sufficient adaptive flexibility, therefore, the P/E-concept predicts different optimum efficiencies of energy conversion according to the selective value of a specific mode of performance. Power strategy leads to an equal balance of dissipation (net-drive) and conservation (compensative output). Higher economical efficiencies are possible, but are only optimum in the long run when energy saving is more successful than a more powerful, fast, and less efficient performance.

COUPLING OF FLOWS AND FORCES, AND ERGOBOLIC POWER

Cells are ergodynamic structures (Fig. 3) in which total energy flow is functionally related to the coupling of input to output flows and input forces to compensative forces. Besides photosynthesis, the most important process of biological energy transformation is that occuring in catabolism. In the overall process, two coupled half reactions are distinguished, one characterized by a high driving force, that is a strongly exergonic (negative),

Fig. 3. Two-compartmental energy transformation in the catabolic-anabolic energy chain. Metabolic flows (half-cycles) are normalized on the basis of ATP-coupling stoichiometries; S/ATP and p/ATP are substrate consumption and product formation, respectively, per molar ATP-cycle. In the metabolic steady state, the ergobolic cycle of ATP-formation (e) and ATP-utilization (de) is balanced. In the dissipative steady state, the energy sinks (ke, ea, and d) balance the catabolic input. d is the dissipative half-cycle, e.g. in protein-turnover or crossbridge cycling. Mismatched drive-coupling may induce compensation by futile (ad) cycles. Excessive load-coupling induces non-steady state depletion of ATP-equivalent energy stores.

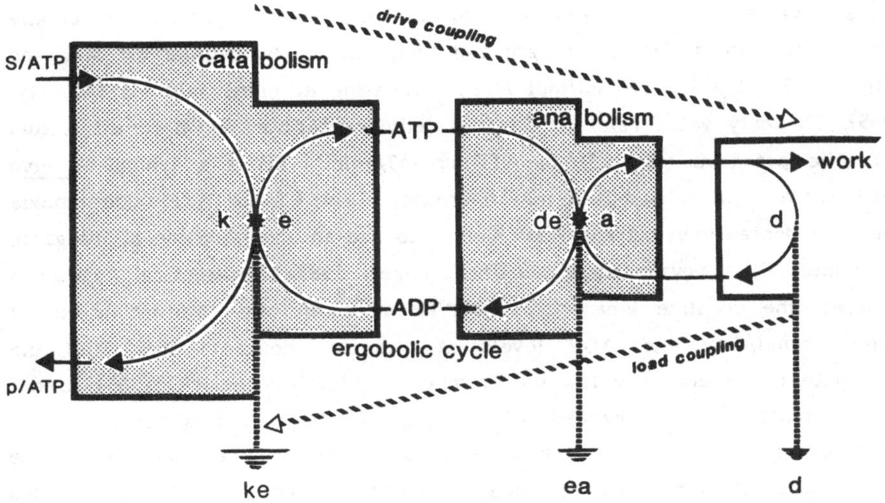

molar Gibbs energy of reaction, and the other by energy conservation (endergonic). Let us define the driving reaction, where reduced fuel substrates (carbohydrate, lipid, protein) are oxidized, as the catabolic half reaction, and the energy-conserving half reaction, where ADP is phosphorylated to ATP and phosphocreatine or phosphoarginine stores are replenished, as the ergobolic reaction.

The term "ergobolic reaction" (Gnaiger, 1983a; compare ergodynamics), indicates that changes in cellular ATP/ADP ratios or phosphagen stores take place by ergodynamic coupling to catabolic or anabolic half reactions (Fig. 3). These ergobolic net changes are endergonic or exergonic, respectively, and have a specific function in regulating the efficiency of metabolic energy transformation. No net changes of adenylates and phosphagens occur in the ergobolic steady state when ATP-turnover mediates the biochemical "energy unit" of ATP from catabolism to anabolism. The ergobolic flow is expressed as the rate of ATP production, ${}_e\dot{N}$ (µmol ATP/h) ($\dot{N}$ denotes dN/dt). The conjugate ergobolic force is the Gibbs energy of phosphorylation of one mole ADP to ATP, that is the phosphorylation potential, $\Delta_e G$:

$$ADP + P_i \rightarrow ATP$$

$$\Delta_e G = \Delta_e G^{o'} + RT \ln \left(\frac{[ATP]}{[ADP] \times [P_i]} \right) \qquad (1)$$

$\Delta_e G^{o'}$ varies as a function of cellular pH, divalent cation (especially magnesium) activity, ionic strength and temperature. At cellular pH of 7 and pMg of 2.7, $\Delta_e G^{o'}$ is 30 kJ/mol ATP, decreasing at more acid pH (Alberty, 1969). The only value relevant for the cellular reactions is $\Delta_e G$, the actual ergobolic potential: +51 kJ/mol ATP or mJ/µmol ATP is a typical in vivo value for aerobic cells, but it may decrease to +44 kJ/mol ATP under anoxia due to a decrease in pH and ATP/ADP-ratio and an accumulation of inorganic phosphate, P_i (Dawson et al., 1978; Gnaiger, 1983a). In resting vertebrate muscles, the creatine kinase buffers ATP at essentially constant levels of some 7mmol/dm^3 and ADP levels at µmolar concentrations, and the phosphate levels are very low (di Prampero, 1981; Kushmerick, 1985). Under these conditions, $\Delta_e G$ exceeds +60 kJ/mol ATP (Eq. 1). This wide range of Gibbs energy changes of phosphorylation is in remarkable contrast to the many textbook statements referring to some constant or cellular standard value.

Catabolic flow, ${}_k\dot{N}_i$, is measured in various ways; as catabolic substrate consumption, ${}_k\dot{N}_s$; as the rate of oxygen consumption, ${}_k\dot{N}_{O2}$; or as the rate of accumulation and excretion of endproducts, ${}_k\dot{N}_p$. Accordingly, the catabolic input force is expressed as the Gibbs energy change per mol substrate ($\Delta_k G_s$ = -2879 or -2873 kJ/mol glycogen, at a glycogen concentration of 1.0 and 0.09 mol/dm^3, respectively, and otherwise typically aerobic cellular conditions; see Gnaiger, 1983a; here glycogen is always used for glycosyl-units). Since 6 mol O_2/mol glycogen are consumed in aerobic catabolism, the catabolic force of oxygen is $\Delta_k G_{O2}$ = -2873/6 = -479 kJ/mol O_2 . This value is nearly identical with the oxycaloric equivalent, $\Delta_k H_{O2}$, for the substrate glycogen (Gnaiger, 1983b). Therefore, the calorimetric measurement of aerobic heat dissipation yields the catabolic power directly (Gnaiger, 1983c). This is not true, however, for anaerobic catabolism.

Based on catabolic pathways or biochemical maps, a specific stoichiometry is calculated between catabolic and ergobolic rates. According to this "mechanistic stoichiometry", we expect 37 (or 38) mol ATP generated per mol glycogen oxidized, which converts to a stoichiometric coefficient for oxygen of 37/6 = 6.17 ATP/O_2. For anoxic catabolism, the mechanistic stoichiometry is 1.5 ATP/lactate and 3 ATP/glycogen, or 2.75 ATP/succinate and 4.71 ATP/glycogen (Gnaiger, 1977; Tab. 1).

The mechanistic molar stoichiometry, $\nu_{ATP/i}$ (i = catabolic reactant), provides the important link between ergobolic and catabolic flows and forces. The catabolic force efficiency, ${}_k f$, and flow efficiency, ${}_k j$, are defined here as the ergobolic/catabolic flow and force ratio, respectively, normalized for the ATP-stoichiometry.

$$ {}_k f = \frac{-\Delta_e G}{(\Delta_k G_i \,/\, \nu_{ATP/i})} \tag{2.1} $$

$$ {}_k j = \frac{{}_e\dot{N}}{({}_k\dot{N}_i \times \nu_{ATP/i})} \tag{2.2} $$

In Eq. (2), the two expressions in brackets can be understood as the catabolic coupling force and flow, respectively, normalized for the mechanistic ATP-coupling stoichiometry. We define (Gnaiger, 1983a)

$$ \Delta_k G = \Delta_k G_i \,/\, \nu_{ATP/i} \tag{3.1} $$

$$ {}_k\dot{N} = {}_k\dot{N}_i \times \nu_{ATP/i} \tag{3.2} $$

From Eq. (2.2) and (3.2) we see that for the fully coupled process (${}_k j$ = 1), the ergobolic rate of ATP production, ${}_e\dot{N}$ (µmol ATP/h) is numerically identical to $_k\dot{N}$, which then is the rate of ATP-turnover (µmol∞ATP/h). In most cases, Eq. (3.2) is the only means for calculating the rate of ATP-turnover, with the assumption that $_kj = 1$.

Using Eq. (2) the force and flow efficiencies are calculated (Tab. 1)

$$_kf = \frac{-\Delta_e G}{\Delta_k G} \qquad (4.1)$$

$$_kj = \frac{_e\dot{N}}{_k\dot{N}} \qquad (4.2)$$

For the above example of glycogen respiration, the aerobic catabolic force efficiency is obtained from Eq. (2.1) or (4.1),

$$_kf = \frac{51}{479/6.17} = \frac{51}{78} = 0.66$$

The product of flow efficiency and force efficiency is the input-output power efficiency or ergodynamic efficiency, η, which equals the force efficiency at $_kj = 1$;

$$_k\eta = {_kj} \times {_kf} = \frac{-_eP}{_kP} \qquad (5)$$

The ergoblic power output, $_eP$ (mJ/h; = mJ/s = mW), is the product of the rate of ATP-formation and ergobolic potential (Eq. A3.2). This composite function of catabolic energy conversion is clearly recognized in biochemical studies where the ATP-equivalent catabolic flux and the ATP/ADP-ratio, pH, etc. are taken into consideration. At low efficiencies, catabolic power input, $_kP$, is mainly dissipated, rendering the ergobolic power output low (Fig. 2; inefficiency). At high efficiency, the net catabolic-ergobolic potential or "ATP-coupling potential" (Eq. A1) is low; therefore the rate of energy conversion and hence power output is low again (Fig. 2; ergodynamic inhibition). The P- and E-optimum functions based on linear ergodynamic equations are derived in the Appendix (Eqs. A4.2 and A5).

LOWEST ATP GAIN - HIGHEST ATP PRODUCTION

During burst activity over periods of some 40 seconds, strenuous muscular exercise is powered by the lactate pathway in vertebrates (di Prampero, 1981). This is also true for invertebrates, where the lactate and the opine pathways (Livingstone et al., 1983) have two features in common, typifying the state of physiological anoxia: (1) these pathways are employed when ATP demand is highest, and (2) they are the pathways with the least gain in ATP per unit substrate, that is 3 ATP/glycogen (Tab. 1). This interconnexion may at first sight appear to be paradoxical: Could an athlete runner win the race by extracting more ATP from the glycogen stores of his muscles? The ergodynamic analysis yields a definitive answer: The high ATP/glycogen coefficients of the succinate-propionate-acetate pathways are unsuitable for fostering high rates and provide relatively low power output, due to ergodynamic inhibition (Fig. 2). Efficiencies exceeding 0.7 are certainly incompatible with P-strategy. This explains why even invertebrates capable of utilizing high-efficiency pathways at low rates, switch to the less efficient lactate pathway to power muscular exercise in the presence and absence of environmental oxygen (Zebe et al., 1981, Putzer, 1985).

Under environmental anoxia, most euryoxic invertebrates produce succinate as the major glycolytic end product initially (I), whereas the propionate-acetate pathway predominates after a transient period (II). The

Table 1. ATP-stoichiometry and ergodynamic characteristics of fully coupled catabolism of glycogen in aerobic muscle (ox) and in anoxic pathways with the formation of different end products (anox). f_{opt} is the theoretical optimum compartmental force efficiency corresponding to the respective strategy in a two-compartmental energy chain. For further explanation see text and Gnaiger (1983a).

Pathway	ATP/glycogen-units		$\Delta_k G$	$\Delta_e G$	$_k f$	function	strategy
	$\nu_{ATP/S}$	%	(kJ/mol ATP)				(f_{opt})
ox	37	100	-78	62	0.80	muscle, rest	E_{II} (0.80)
ox			-78	50	0.64	muscle, high ox exercise	P (0.67)
lactate	3.0	8.1	-81	51	0.63	muscle, onset of anox exercise	P (0.67)
succinate	4.71	12.7	-70	51	0.74	environ. anox, initial	E_I (0.75)
prop.-acetate	6.33	17.1	-56	44	0.79	environ. anox, long-term	E_{II} (0.80)
propionate	6.43	17.4	-55	44	0.81	environ. anox, long-term	E_{II} (0.80)

stoichiometric ATP coupling coefficient, $\nu_{ATP/S}$(the "biochemical efficiency"), of these pathways is high relative to lactate-glycolysis (Tab. 1), and coincides with a decreased proton generation per mol ATP-turnover (Gnaiger, 1980a; Pörtner et al., 1984). These traits are generally recognized as the most significant biochemical adaptations to anoxic tolerance. However, the biochemical-stoichiometric perspective does not reveal any advantage of initial succinate accumulation over propionate-acetate production. During aerobic recovery, propionate is resynthesized to glycogen as readily as is succinate. Yet, the great majority of euryoxic invertebrates accumulate and excrete propionate only after an initial lag-time which may last up to 18 hours (Kluytmans et al., 1978). Is it lack of enzyme-kinetic flexibility, that prevents an immediate metabolic switch to the propionate pathway with a yield of 6.4 mol ATP/mol glycogen instead of only 4.7 mol ATP/mol glycogen in the succinate pathway?

An ergodynamic analysis is required to understand the adaptive significance of initial succinate accumulation and secondary propionate-acetate production, and to fully appreciate the kinetic fine-tuning of the metabolic machinery. An economical energy converter must not only transform free energy at a characteristic optimum efficiency but must, first of all, extract effectively free energy from the available substrate ($\Delta_k G \times \nu_{ATP/S}$ must be high; Tab. 1). A more detailed analysis of the complex compromise in optimization of catabolic energy conversion helps to explain why and when the succinate or the propionate-acetate pathway are most advantageous (Gnaiger, 1983a).

However, three questions have yet to be solved: (1) How can we rationalize the high force efficiencies observed in aerobic and anaerobic catabolism? If economical power output in terms of ATP were the definitive fitness parameter in E-strategy, then the force efficiency is expected to be 0.67, instead of 0.74 and 0.80 for succinate (I) and propionate (II) respectively. Moreover, the lactate pathway - the most typical example for P-strategy - should operate at a force efficiency of 0.5 instead of >0.6, if the system were selected for maximum ergobolic power output. (2) What is the rate of anoxic ATP-turnover? A discrepancy between anoxic heat dissipation and enthalpy changes of biochemical reactions suggests that ATP-turnover, calculated from biochemical measurements, presently underestimates total catabolic rates under anoxia (Gnaiger, 1980b; 1983a; Shick et al., 1983). (3) What is the functional significance of ATP production? Metabolism is not simply designed to produce ATP. ATP is an intermediate for transmission of energy to various ATP-utilizing processes. ATP-turnover is the energetic coupling of catabolism and anabolism; hence the term "ergobolic cycle" (Fig. 3). Energy metabolism can be understood only as the integrated process of anabolic ATP utilization and catabolic ATP formation. This argument is simple but important, yet it is frequently ignored.

The catabolic-ergobolic optimum efficiencies in P- and E-strategy in one compartment do not retain their characteristic values when seen in the context of whole-system function. Analysis of the metabolic system as a two-compartmental energy chain (Fig. 3) explains the high catabolic efficiency as optimization of energy-chain efficiency.

THE TWO-COMPARTMENTAL CATABOLIC-ANABOLIC ENERGY CHAIN

Metabolism is structured into catabolic and anabolic reaction sequences (Fig. 3). This partitioning is a prerequisite for the flexibility and regulatory capacity of cellular metabolism to utilize various fuel substrates for the entire variety of "anabolic" functions such as biosynthesis, active transport, electrical signal transmission and mechanical work, via ATP (Atkinson, 1977). For metabolic control, the conductance of catabolism must be matched with that of anabolism, and this in turn must be adjusted to the external load conductance (Appendix A.2). Instead of catabolic efficiency and ergobolic power output, now the overall efficiency of the energy chain and the anabolic power output are recognized as the proper criteria for optimum cellular functions.

The anabolic power output is here defined as the power of any process that is driven by coupling with the utilization of ATP-equivalent energy. The anabolic coupling force, $\Delta_a G$, is the output force normalized in relation to the ergobolic driving force, $-\Delta_e G$ (compare Eq. 2 to 4). For instance, $\Delta_a G$ is the force generated in an actin-myosin crossbridge that is formed in one ATP hydrolysis cycle. Analogous to the efficiency of the catabolic compartment, the anabolic force efficiency is defined as

$$_a f = \frac{\Delta_a G}{\Delta_e G} \quad (6)$$

The overall- (ka-) efficiency of the two-compartmental energy chain, $_{ka}f$, is the product of the catabolic and anabolic force efficiencies (Wilkie, 1974),

$$_{ka}f = {_k f}\ {_a f} \quad (7)$$

If any one of these efficiencies is near zero, then the anabolic power output is near zero. Conversely, if one efficiency is unity, then the flow through the respective compartment is regulated entirely by the other energy converter; the two compartments merge effectively into one.

This ergodynamic definition of compartmentalization is amply reflected by local separation of catabolic and anabolic functions in cell organelles and tissues. Differentiation of cytosolic and mitochondrial potentials of ATP, yields information on different control sites of metabolic flux (Klingenberg & Heldt, 1982), as does the distinction of reversible and irreversible enzyme steps within a metabolic reaction sequence (Newsholme & Start, 1973). The following consideration is restricted to ergodynamically defined compartments of energy transformation and to bulk estimations of chemical potentials (Tab. 1).

Enzymes and membrane transport systems exert "active" control over the conductance coefficients of catabolism and anabolism (${}_kL$ and ${}_aL$, the kinetic constants; see Appendix). Like the ATP-stoichiometry, the conductance coefficients are evolved features of the metabolic mechanisms maintaining living systems in a functional state. One strong argument is that, for reasons of regulatory capacity, both energy transforming compartments are optimized according to P- or E-strategy. In the steady state of such systems, the catabolic and anabolic conductance coefficients, ${}_kL$ and ${}_aL$, must be tuned such that the internal conductance ratio (Eq. A8), ${}_{ka}m = {}_kL/{}_aL$, equals the compartmental (catabolic or anabolic) force efficiency. This specific stability criterion is derived in Appendix A.3 (for ${}_kf = {}_af$). In order to increase the ergobolic potential (Eq. 1) for obtaining a high economical catabolic efficiency, the anabolic load or conductance for ATP-utilization, ${}_aL$, must decrease (${}_{ka}m$ increases; E-strategy). Then a state of "drive-coupling" is maintained (Fig. 3). Conversely, the ergobolic potential drops in response to a lowering of anabolic force efficiency and increased anabolic load, ${}_aL$ (${}_{ka}m$ decreases), whence the catabolic force efficiency decreases (P-strategy) and the rate of ATP-regeneration speeds up (Eq. A2). This mechanism maintains then a new steady-state in response to a step-increase of the external load ("load-coupling"; Fig. 3). This ergodynamic model incorporates active kinetic control of catabolic and anabolic conductance coefficients. It is consistent with the view (Thauer <u>et al.</u>, 1977) that fast aerobic growth is controlled by anabolism (load-coupling in P-strategy), whereas catabolism appears to be rate limiting in many anaerobes (drive-coupling in E-strategy; Fig. 3).

Consideration of P- or E-strategy with respect to compartmental energy chains suggests a serious problem in traditional interpretations of optimum efficiencies. If one compartment of a two-compartmental system is singled out (oxidative phosphorylation; Stucki, 1980), then a catabolic efficiency of 0.67 would be interpreted as indicating E-strategy on the basis of one-compartmental analysis of a fully coupled process (Fig. 2a). However, the ka-efficiency of the entire energy chain would then amount to 0.44, taking hypothetically equal compartmental efficiencies as an example (Fig. 4.1; ${}_{k}f = {}_{a}f$). Certainly, a system displaying an input-output force efficiency of <0.5 can no longer be interpreted as E-strategy. On the other hand, the catabolic-anabolic energy chain of bacterial growth has been subjected to a black-box analysis (Westerhof et al., 1983). Consequently, these authors interpreted a total force efficiency of $\sqrt{0.67}$ as E-strategy. However, the compartmental force efficiency would then have to be 0.67 = 0.82 (Eq. A13), far beyond previous expectations for simple E-strategy.

With the concept of the ergodynamic structure of the catabolic-anabolic energy chain, it is possible to explain the high compartmental force efficiencies of anoxic catabolism. The optimum efficiency for maximum anabolic power output turns out to be ${}_{k}f = {}_{a}f = 0.67$ with a ka-force efficiency of 0.44 (Fig. 4, P). While 0.67 is the E-optimum in one-compartmental systems (Fig. 2), the same characteristic value is now recognized as the P-optimum for a two-compartmental system. This prediction is consistent with the expectation that the lactate pathway evolved under the selective pressure favouring P- rather than E-strategy to power hypoxic activity (Sidell & Beland, 1979; Tab. 1). Muscular power output provides a means of measuring external transmission of anabolic power directly. The maximum efficiency of the anabolic compartment is 0.66 in DNFB-poisoned muscle (dinitrofluorobenzene inhibits phosphocreatine splitting; Kushmerick & Davies, 1969; Wilkie, 1974). This supports the assumption that catabolic and anabolic efficiencies are tuned at the same level according to P-strategy (Fig. 4, P). Moreover, in high aerobic exercise the maximum overall ka-efficiency in large vertebrates is 0.41 (Heglund & Cavagna, 1985; compare ${}_{ka}f_{opt}$ = 0.44; Tab. 2). Reference to experimental maximum efficiencies of mechanical power output is warranted when these external ergodynamic efficiencies are taken as indications of internal force efficiencies (see Eq. 5). At the extreme of zero external efficiency at isometric contraction, external transduction of anabolic force is prevented. This apparent decoupling effect is minimized at maximum external efficiencies.

E-strategy of a two-compartmental system predicts an optimum

Fig. 4. Optimization of catabolic force efficiency in catabolic-anabolic coupled metabolism at constant catabolic input force. Two linear systems (Fig. 2a) are connected in series. ${}_{ka}f$ is the catabolic-anabolic force efficiency of the system under the assumption that both compartments operate at identical efficiencies. k, e, and a are catabolic power input, ergobolic and anabolic power output, respectively, normalized relative to maximum power input in P-strategy (4.1-4.3).

4.1 P-strategy: The optimum catabolic force efficiency for ${}_eP$ is 0.50 (Fig. 2a) but 0.67 for maximum anabolic power output (Eq. A14, m=2).

4.2 E_I-strategy: The optimum force efficiency for phase-I economical ${}_eP$ is 0.67 (Fig. 2a) but 0.75 for maximum economical ${}_aP$ (Eq. A14, m=3).

4.3 E_{II}-strategy: The product of efficiencies in both compartments defines the degree of phase-II economy. Optimum efficiencies for ${}_eP$ and ${}_aP$ are 0.75 and 0.80 respectively (Eq. A14, m=4).

4.4 Comparison of the optimum functions of two-compartmental steady-state anabolic power output (from 4.1-4.3, normalized at unit ${}_kL$) and one-compartmental transient-state anabolic power output (P -strategy, normalised relative to maximum power output at unit ${}_aL$; Eq. A14, m=1). The continuous line intersecting the power maxima is a plot of Eq. A14 with m substituted from Eq. A10. This plot can be interpreted as the optimized power output of a non-linear system with f^{m-2} as feedback parameter (and hence omitted from the left side of Eq. A14). In the physiological range of force efficiencies from 0.5 to 0.8, an apparent linear relationship is then observed between power output and force efficiency.

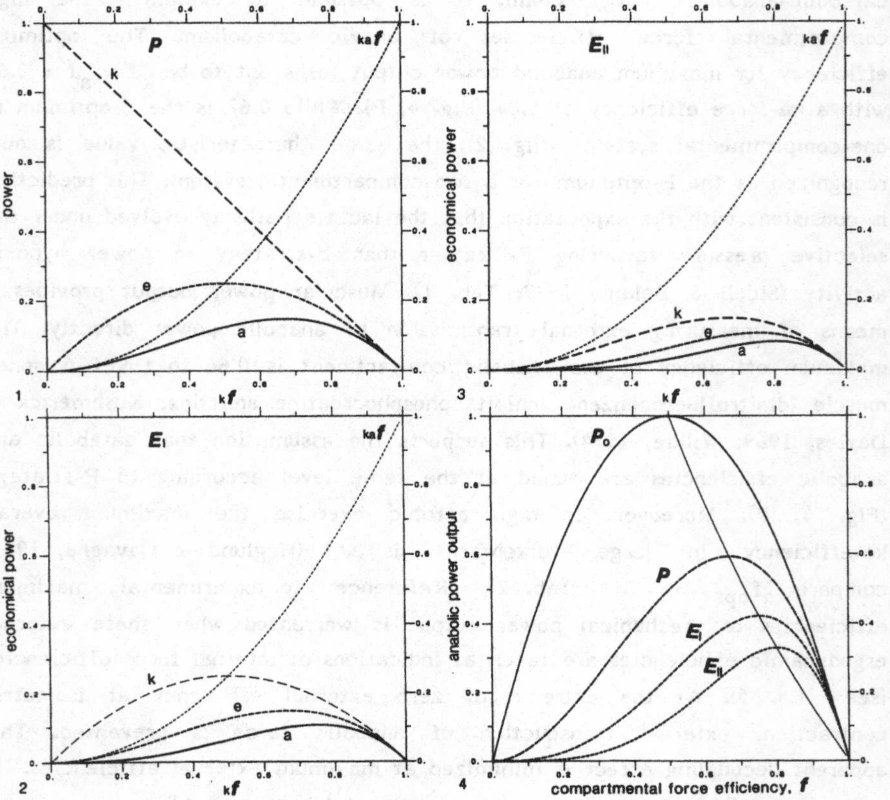

compartmental efficiency of 0.75 (Tab. 2; Fig. 4, E_I). The force efficiency for the succinate pathway of 0.74 appears no longer to be unreasonably high (Tab. 1).

In a two-compartmental system, an additional degree of freedom is gained in the development of E-strategy: A second stage of economy can be expected when the efficiency of both compartments counts as a weighting factor in the fitness parameter (Appendix A.3). Consequently, for E_{II}-strategy an optimum compartmental force efficiency of 0.80 is predicted (Fig. 4, E_{II}). Actually, this agrees with the catabolic force efficiency of the propionate and propionate-acetate pathway in the secondary phase of long-term anoxia (Tab. 1).

The theoretical definition of distinctive strategies does not necessarily imply discontinuous jumps between P- and E-optimum states. While the switching on and off of a biochemical pathway actually represents a step change in the normalized catabolic input force, the distinct strategies overlap due to gradual changes of ergobolic potentials. Accordingly, a continuous ergodynamic optimum function (Eq. A14) describes the intermittent optimum efficiencies of a fine-tuned energy chain in the steady state (Fig. 4.4). In the next section this analysis is extended to highlight the ergodynamic significance of a prominent non-steady state process in muscular physiology.

Table 2.

P- and E_I-optimum force efficiencies in a two-compartmental energy chain. The optimum efficiencies for one compartment, ${}_kf$, are higher, and the optima for the whole energy chain, ${}_{ka}f$, are lower than the optimum efficiencies in a one-compartmental system (underlined numbers). P and E_I are given in per cent of their maximum values at the respective optimum efficiencies of the energy chain.

${}_kf$	${}_{ka}f$	P	${}_kf$	${}_{ka}f$	E_I
<u>0.50</u>	0.25	84.4	<u>0.667</u>	0.444	93.6
0.667	0.444	100	0.75	0.563	100
0.707	<u>0.50</u>	98.9	0.816	<u>0.667</u>	94.7

THE DUAL ROLE OF PHOSPHOCREATINE IN P- AND E-STRATEGY

The predominant role of phosphocreatine or phosphoarginine in powering maximum activity during short periods of time is widely documented ("all-out efforts"; di Prampero, 1981) and is recognized as a non-steady-state process. The creatine kinase reaction is in equilibrium with the ATP-system, so ergobolic depletion by maximum ATP-demand conforms to a one-compartmental energy transformation. In this state of load-coupling (Fig. 3), the anabolic conductance for ATP-utilization far exceeds the limiting condition of steady-state conductance matching (Eq. A8). This is the kinetic advantage of a one-compartmental strategy for maximizing instantaneous anabolic power output (P_0-strategy, Fig. 4.4).

P_0-strategy for one completely coupled compartment determines the optimum force efficiency at 0.50. This is also the experimentally observed efficiency of mechanical transduction of phosphocreatine splitting (Curtin <u>et al</u>., 1974). The internal efficiency loss of a series of energy converters is avoided in one-compartmental transformation. If two energy converters are connected in series and each operates at 50% efficiency, then the overall efficiency of the system reduces to 25% (Eq. 7). In that mode of operation, however, anabolic power output would be 16% below the two-compartmental maximum (Tab. 2). Clearly, a price must be paid for the advantage of regulatory capacity obtained in compartmental, catabolic-anabolic energy conversion. The one-compartmental, transient-state mechanism of phosphocreatine or phosphoarginine depletion provides the more powerful strategy in explosive all-out efforts. Due to its simplicity, it may well be the most primitive mechanism of biological energy coupling (Koch, 1986), originally evolved without constraints set by energy resources and clearly in a situation favouring r-strategy.

In addition to this anoxic involvement of creatine kinase, its integration in aerobic ATP transport, and in buffering ATP and ADP concentrations is well recognized (Hochachka & Somero, 1984; Stucki, 1980). A regulatory function of the creatine kinase reaction in aerobic muscle exercise has only recently been suggested (Kushmerick, 1985). The ergodynamic interpretation of these data indicates a transition from E-strategy to P-strategy as the muscle switches from a resting state to high aerobic exercise.

Application of the concept of a fully coupled energy chain to interpret aerobic metabolism requires a comment at first. One-compartmental black-box analysis and emphasis on phenomenological coupling coefficients (Caplan & Essig, 1983) may mask some functional relationships of catabolic rate and force efficiency. On a whole-organism or cellular level, the significance of the standard metabolic rate is readily accepted in terms of maintenance requirements (Bayne, 1986) or spinning of futile cycles (Hue, 1982). Why should any maintenance requirements be absent on the subcellular level? Since mitochondrial state-4 respiration and coupling coefficients <1.0 may, to a large extent, be related to mitochondrial "anabolic" ATP-requirements for maintenance, phenomenological decoupling cannot be interpreted as functional decoupling. Instead of invoking the hypothesis of maximization of efficiency by decoupling (Stucki, 1980), force efficiencies of oxidative phosphorylation can be interpreted within the framework of a fully coupled (Harris et al., 1980), two-compartmental energy chain (this does not, however, exclude a completion of the dissipative cycle, as illustrated in Fig. 3). A specific bypass mechanism (site I in yeast; Erecinska et al., 1978) as opposed to decoupling influences the force efficiency via stoichiometry; a bypass may optimize the force efficiency comparable to the switching between different anoxic pathways. The involvement of phosphocreatine breakdown in the transition towards increasing aerobic work loads can be interpreted as an important ergodynamic control mechanism in the energy chain (a more detailed explanation: Gnaiger, in prep).

From NMR- (nuclear magnetic resonance) studies on vertebrate muscle operating at different aerobic steady-state levels (Kushmerick, 1985), it can be calculated that the efficiency of aerobic catabolism decreases from a resting level of 0.80, to 0.64 at high aerobic exercise (Tab. 1; ox). This change of catabolic force efficiency is induced by a decrease of phosphocreatine content and an increase in inorganic phosphate and ADP concentrations with increasing aerobic work load (Kushmerick, 1985). Therefore, oxidative phosphorylation operates at the E_{II}-optimum at rest (compare ${}_{k}f = 0.79$ for oxidative phosphorylation in liver in a "resting state"; Stucki, 1980), passes through the E_{I}-optimum at low activity levels, and operates according to P-strategy at high steady-state exercise (Fig. 4). Appropriate tuning of anabolic force efficiencies is possible by adjusting the sliding distance of the actin filaments per mol ATP-turnover (Yanagida et al., 1985).

GENERAL CONCLUSIONS

Interactive tuning of catabolic and anabolic conductance

coefficients is required to stabilize compartmental force efficiencies at optimum levels (Fig. 4). In the life time of organisms, adjustments of conductance coefficients are possible within genetically fixed limits. Random sets of enzymes and other variables of rate control will eventually lead to steady states (Kacser & Burns, 1973), but these would not produce predictable optimum patterns. Ergodynamic optimum functions provide a baseline by which the vast diversity of potentially adaptive enzyme-kinetic traits can be rationalized. Linear or non-linear ergodynamic concepts of metabolic energy chains offer an a priori approach (in the sense of Calow & Townsend, 1981) to the study of biochemical adaptation. This approach may lend more credibility to the expanding theory on metabolic strategies, by removing circular arguments and limiting otherwise virtually endless parades of ad-hoc constructions on adaptation.

In evolutionary time, selection of ATP-coupling stoichiometries and of basic enzyme-kinetic properties have produced economically regulated systems. These are characterized by a synergistic relation between active enzyme kinetic control of metabolic conductance and ergodynamic control via driving forces and compensative coupling. Ergodynamic optimum functions emphasize the complementary nature of active and passive regulation. The theory of two-compartmental optimum functions postulates the matching of catabolic and anabolic conductance coefficients in relation to the load. Over-proportional activity of enzymes in an anabolic pathway yields a lower fitness than intermediate improvements of enzymes in catabolic and anabolic pathways. Evolutionary biologists generally find intermediate enzyme kinetic properties associated with protein heterozygosity, and recognize that "a chain of intermediate fitnesses can result in superior fitness" (Mitton & Grant, 1984). These theoretical predictions are consistent with empirical correlations of heterozygosity, growth efficiency and growth rate (Bayne, 1986). Finally, for translating physiological fitness into evolutionary fitness, the possible advantages of optimization and scope for regulation over maximization of rates must be considered in relation to environmental perturbations (Calow, 1984; Stebbing & Heath, 1984; Wieser, 1985), ecosystem stability (Ott, 1981) and cultural developments (Talsma, 1980).

Without incorporation of evolutionary mechanisms of optimization or adaptation, previous physicochemical attempts to describe biological systems or biological evolution failed to provide an adequate thermodynamic definition of life. The classical approach based on the Second Law or on states of "negentropy" (Schrödinger, 1944) does not address specifically biological features, since local decreases of entropy or increases in "order" are commonplace in inhomogeneous abiotic systems. The theory on "dissipative structures" includes, on the other hand, explicitly the time-evolution and statistical properties of macroscopic systems (Prigogine, 1980). While this theory is important for understanding the spontaneous formation of sometimes rather spectacular structures far from equilibrium, the time-evolution of the dissipative structures relates to an ontogenetic development of non-linear systems but not to biological evolution which requires a sequence of generations of quasi-repetitive systems. Indeed, it would be surprising if biological evolution could be modelled without incorporating the recognized basis of evolutionary theory. Ergodynamic optimum functions provide a link between physicochemical concepts and the biological theory of evolution. As a consequence of restrictions on the variety of possible steady-states by selection, optimized systems survive, and propagate genetically the traits of matched catabolic and anabolic conductance coefficients as well as the scope for fine-tuning of these coefficients. By developing and genetically propagating properties of ergodynamic optimum structures, living systems are invariably distinguished from the non-living world.

APPENDIX

A.1. LINEAR P- AND E-OPTIMUM FUNCTIONS

Based on the theory of linear energy converters (Kedem & Caplan, 1965), the fully coupled rate of ATP production is related to the net catabolic-ergobolic potential or "ATP-coupling potential", $\Delta_{ke}G$,

$$_e\dot{N} = -_kL\ (\Delta_kG + \Delta_eG) = -_kL\ \Delta_{ke}G \tag{A1}$$

where $_kL \geq 0$ is the phenomenological conductance for the catabolic process. The condition of complete coupling implies $_e\dot{N} = {}_k\dot{N}$ (Eq. 4.2). Substituting for Δ_eG from Eq. (4.1) yields the relationship between ergobolic flow and force efficiency,

$$_e\dot{N} = -_kL\ \ \Delta_k\dot{G}\ \ (1 - {}_kf) \tag{A2}$$

Power is the product of a flow times the conjugate force,

$$\text{input:}\quad _kP = {}_k\dot{N}\ \ \Delta_kG \tag{A3.1}$$

output: $${}_{e}P = {}_{e}\dot{N}\ \Delta_{e}G \qquad \text{(A3.2)}$$

The catabolic power input, ${}_{k}P$, and ergobolic power output, ${}_{e}P$, are a quadratic function of the catabolic coupling force. Combining Eqs. (4.1), (A2) and (A3) (under the condition of Onsager symmetry, see Caplan & Essig, 1983),

$${}_{k}P = -{}_{k}L\ \Delta_{k}G^{2}\ (1 - {}_{k}f) \qquad \text{(A4.1)}$$

$${}_{e}P = {}_{k}L\ \Delta_{k}G^{2}\ {}_{k}f\ (1 - {}_{k}f) \qquad \text{(A4.2)}$$

${}_{e}P$ is maximum for any fixed catabolic coupling force in Eq. (A4.2) if ${}_{k}f$ = 0.50 (P-strategy; Fig. 2). One characteristic degree of economical power output (Stucki, 1980),

$${}_{e}P\ {}_{k}f = {}_{k}L\ \Delta_{k}G^{2}\ {}_{k}f^{2}\ (1 - {}_{k}f) \qquad \text{(A5)}$$

is maximum for any fixed catabolic force, $\Delta_{k}G$ in Eq. (A5), if ${}_{k}f$ = 0.667 (E-strategy, Fig. 2).

A.2. CATABOLIC-ANABOLIC COUPLING. THE STEADY STATE

To maintain metabolic control, the catabolic and anabolic compartment must operate at a steady-state, i.e. input and output flows for the energy converters must be equal. Analogous to catabolic input (Eq. A2), in a fully coupled system anabolic ATP input, ${}_{de}\dot{N}$, is controlled by the anabolic force efficiency and conductance coefficient, ${}_{a}L \geq 0$.

$${}_{de}\dot{N} = {}_{a}L\ \Delta_{e}G\ (1 - {}_{a}f) \qquad \text{(A6)}$$

Substituting for $\Delta_{e}G$ from Eq. (4.1) yields

$${}_{de}\dot{N} = -{}_{a}L\ \Delta_{k}G\ {}_{k}f\ (1 - {}_{a}f) \qquad \text{(A7)}$$

At steady state, catabolic output rate (Eq. A2) equals anabolic input (Eq. A7). This equality yields the internal steady-state conductance ratio, ${}_{ka}m$,

$${}_{ka}m = \frac{{}_{k}L}{{}_{a}L} = {}_{k}f\ \frac{1 - {}_{a}f}{1 - {}_{k}f} \qquad \text{(A8)}$$

The normalized external flow of the anabolic product, expressed as ATP-equivalents (compare Eq. 3.2), is a linear function of the anabolic output force and the external (load) conductance, ${}_{ex}L$,

$$ {}_{ex}\dot{N} = {}_{ex}L \; \Delta_e G \; {}_a f \tag{A9} $$

Again, expression (A9) must equal the anabolic flow (Eq. A6) to ensure steady-state conditions. This requires that the external steady-state conductance ratio, m, is tuned at

$$ m = \frac{{}_a L}{{}_{ex}L} = \frac{{}_a f}{1 - {}_a f} \tag{A10} $$

By multiplying the anabolic input flow (Eq. A6) times the anabolic input force, $-\Delta_e G$, we obtain the anabolic power input (Eq. A11.1); the anabolic power input times the force efficiency yields in turn the anabolic power output in the case of complete coupling (Eq. A11.2),

$$ {}_{de}P = -{}_a L \; \Delta_e G^2 \; (1 - {}_a f) \tag{A11.1} $$

$$ {}_a P = {}_a L \; \Delta_e G^2 \; {}_a f \; (1 - {}_a f) \tag{A11.2} $$

Finally, after substituting in Eq. (A11.2) for ${}_a L$ and $\Delta_e G$ from Eq. (A8) and (4.1) respectively, the anabolic power output is

$$ {}_a P = {}_k L \; \Delta_k G^2 \; {}_k f \; {}_a f \; (1 - {}_k f) \tag{A12} $$

A.3. THE TWO-COMPARTMENTAL OPTIMUM EFFICIENCY FUNCTION

If catabolic and anabolic force efficiency are equal, then the ka-efficiency is a quadratic function of the compartmental efficiency, f (Fig. 4; Eq. 7),

$$ {}_{ka}f = f^2 \tag{A13} $$

For such a system, the anabolic power output (Eq. A12) can be rewritten as

$$ {}_a P \; f^{m-2} = {}_k L \; \Delta_k G^2 \; f^m (1 - f) \tag{A14} $$

In Eq. (A14) an efficiency parameter with the exponent m was introduced which is related to P- and E-strategy. m = 2 is obtained in the straightforward derivation of Eq. (A14) from (A12), which indicates power-strategy (Fig. 4; P). m = 3 is derived from Eq. (A5) for economical power output in one compartment (Fig. 4; E_I). A second economy stage can be distinguished at m=4 (Fig. 4; E_{II}). For a two-compartmental energy chain this function can be rationalized as indicating economy strategy for both energy transforming compartments; then the more pronounced E_{II}-strategy prevails.

Insertion of the characteristic values of m for P-, E_I- or E_{II}-strategy into Eq. (A10) yields the optimum compartmental efficiencies, f_{opt}, at which any chosen fitness parameter attains its maximum value (Fig. 4.4),

$$f_{opt} = \frac{m}{1 + m} \qquad \text{(A15)}$$

Comparison of Eq. (A12) and (A14) shows that for the maintenance of a steady-state the external conductance ratio is restricted to m $\geq$ 2.0. This lower boundary condition for the steady state yields the P-optimum force efficiency of 0.67 for one compartment in the energy chain (P-strategy). Note that a catabolic efficiency of 0.67 would be mistakenly interpreted as indicating E-strategy, if the implications of the two-compartmental energy chain are ignored (Tab. 2).

The physical interpretation of Eq. (A15) is surprising: The external steady state conductance ratio, m (Eq. A10), is the exponent in Eq. (A14). In physical terms, m is the relative endurance time during which f units of energy input provide a constant power, whereas 1+m is the relative time required for the provision of f units of energy output at constant power. This leads to a definition of efficiency as a ratio of time. The present analysis is restricted to fully coupled systems where the efficiency can be equivalently calculated on the basis of force, energy and power ratios. Now efficiency can also be expressed as a time ratio.

$$f_{opt} = \frac{{}_{in}t^*}{{}_{out}t^*} \qquad \text{(A16)}$$

${}_{in}t^*$ = (1+m) t^o is the absolute endurance time in units of t^o during which one unit of energy input drives the coupled process at a constant rate. ${}_{out}t^*$ is the time required to obtain one unit of energy output at a constant rate. P-strategy minimizes the time ${}_{out}t^*$. ${}_{in}t^*$ sets the value of endurance time in relation to power output, and is the quantity required to find the optimum in the compromize between maximum time and maximum power output in E-strategy. This is a non-mechanistic, ergodynamic view of optimum efficiency.

The two-compartmental ergodynamic model as introduced here is applied to rationalize physiological fitness in cellular energy metabolism.

This work was supported by the Fonds zur Förderung der wissenschaftlichen Forschung in Österreich, project J0011. I thank Drs. B.L. Bayne, G. Bitterlich, A. Duncan, A.L. Koch, R.C. Newell and W. Wieser for discussions and constructive comments.

REFERENCES

Atkinson, D.E. (1977) Cellular Energy Metabolism and its Regulation. Academic Press, New York.

Alberty, R.A. (1969) Standard Gibbs free energy, enthalpy and entropy changes as a function of pH and pMg for several reactions involving adenosine phosphates. J.Biol.Chem., 244, 3290-3302.

Bayne, B.L. (1986) Genetic aspects of physiological adaptation in bivalve molluscs. In: Evolutionary Physiological Ecology, Ed. P. Calow. Cambridge Univ. Press, London.

Boyce, M.S. (1984) Restitution of r- and K-selection as a model of density-dependent natural selection. Ann.Rev.Ecol.Syst., 15, 427-447.

Broda, E. (1975) The Evolution of the Bioenergetic Processes. Pergamon Press, Oxford.

Calow, P. (1984) Economics of ontogeny - adaptational aspects. In Evolutionary Ecology, Ed. B. Shorrocks, pp. 81-104. Blackwell Sci. Publ., Oxford.

Calow, P. & Townsend, C.R. (1981) Energetics, ecology and evolution. In Physiological Ecology. An Evolutionary Approach to Resource Use, Eds. C.R. Townsend & P. Calow, pp. 3-19. Blackwell Sci. Publ., Oxford.

Caplan, S.R. & Essig, A. (1983) Bioenergetics and Linear Nonequilibrium Thermodynamics. The Steady State. Harvard University Press, Cambridge, Massachusetts.

Crawford, V. (1978) Biochemical and ecological similarities in marsh plants and diving animals. Naturwiss., 65, 194-201.

Curtin, N.A. et al. (1974) The effect of the performance of work on total energy output and metabolism during muscular contraction. J. Physiol., 238, 455-472.

Dawson, M.J. et al. (1978) Muscular fatigue investigated by phosphorus nuclear magnetic resonance. Nature, 274, 861-866.

Erecinska, M. et al. (1978) Homeostatic regulation of cellular energy metabolism: experimental characterisation in vivo and fit to a model. Am. J. Physiol., 234, C82-C89.

Fenchel, T. & Blackburn, T.H. (1979) Bacteria and Mineral Cycling. Academic Press, London.

Gnaiger, E. (1977) Thermodynamic considerations of invertebrate anoxibiosis. In: Application of Calorimetry in Life Sciences, Eds. I. Lamprecht & B. Schaarschmidt, pp. 281-303. de Gruyter, Berlin.

Gnaiger, E. (1980a) Das kalorische Äquivalent des ATP-Umsatzes im aeroben und anoxischen Metabolismus. Thermochim. Acta, 40, 195-223.

Gnaiger, E. (1980b) Energetics of invertebrate anoxibiosis: direct calorimetry in aquatic oligochaetes. FEBS Lett., 112, 239-242.

Gnaiger, E. (1983a) Heat dissipation and energetic efficiency in animal anoxibiosis: economy contra power. J. Exp. Zool., 228, 471-490.

Gnaiger, E. (1983b) Calculation of energetic and biochemical equivalents of respiratory oxygen consumption. In: Polarographic Oxygen Sensors. Aquatic and Physiological Applications, Eds. E. Gnaiger & H. Forstner, pp. 337-345. Springer, Berlin, Heidelberg, New York.

Gnaiger, E. (1983c) The twin-flow microrespirometer and simultaneous calorimetry. In: Polarographic Oxygen Sensors. Aquatic and Physiological Applications, Eds. E. Gnaiger & H. Forstner, pp. 134-166. Springer, Berlin.

Gnaiger, E. (in prep.) The Power of Life.

Harris, S.I. et al. (1980) Oxygen consumption and cellular ion transport: Evidence for adenosine triphosphate to O_2 ratio near 6 in intact cell. Science, 208, 1148-1150.

Heglund, N.C. & Cavagna, G.A. (1985) Efficiency of vertebrate locomotory muscles. J. Exp. Biol., 115, 283-292.

Hildebrandt, S. & Tromba, A. (1985) Mathematics and Optimal Form. Sci. American Library, New York.

Hochachka, P.W. & Somero, G.N. (1984) Biochemical Adaptation. Princeton, Univ. Press, Princeton, New Jersey.

Hue, L. (1982) Futile cycles and regulation of metabolism. In: Metabolic Compartmentation, Ed. H. Sies, pp. 71-97. Academic Press, London.

Kacser, H. & Burns, J.A. (1973) The control of flux. Symp. Soc. Exp. Biol., 32, 65-104.

Kedem, O. & Caplan, S.R. (1965) Degree of coupling and its relation to efficiency of energy conversion. Trans. Faraday Soc., 61, 1897-1911.

Klingenberg, M. & Heldt, H.W. (1982) The ADP/ATP translocation in mitochondria and its role in intracellular compartmentation. In: Metabolic Compartmentation, Ed. H. Sies, pp. 101-122. Academic Press, London & New York.

Kluytmans, J.H. et al. (1978) Production and excretion of volatile fatty acids in the sea mussel Mytilus edulis L. J. Comp. Physiol., 123, 163-167.

Koch, A.L. (1985) The macroeconomics of bacterial growth. In: Bacteria in their Natural Environment, Eds. M.M. Fletcher & G.D. Floodgate. Symp. Soc. General Microbiol.

Koch, A.L. (1986) Evolution from the viewpoint of Escherichia coli. In: Evolutionary Physiological Ecology, Ed. P. Calow, Cambridge Univ. Press, London.

Krebs, J.R. & McCleery, R.H. (1984) Optimization in behavioural ecology. In: Behavioural Ecology. An Evolutionary Approach, Eds. J.R. Krebs & N.B. Davies, pp. 91-121. Blackwell Sci. Publ., Oxford.

Kuhn, T.S. (1970) The Structure of Scientific Revolutions, 2nd edn. Chicago Univ. Press, Chicago.

Kushmerick, M.J. (1985) Patterns in mammalian muscle energetics. J. Exp. Biol., 115, 165-177.

Kushmerick, M.J. & Davies, R.E. (1969) The chemical energetics of muscle contractions. II. The chemistry, efficiency and power of maximally working sartorius muscles. Proc. R. Soc., B 174, 315-353.

Livingstone, D.R. et al. (1983) Studies on the phylogenetic distribution of pyruvate oxidoreductases. Biochem. Syst. Ecol., 11, 415-425.

MacArthur, R.H. & Wilson, E.O. (1967) The Theory of Island Biogeography. Princeton Univ. Press, Princeton.

Maynard Smith, J. (1978) Optimization theory in evolution. Ann. Rev. Ecol. Syst., 9, 31-56

Mitton, J.B. & Grant, M.C. (1984) Associations among protein heterozygosity, growth rate and developmental homeostasis. Ann. Rev. Ecol. Syst., 15, 427-447.

Newsholme, E.A. & Start, C. (1973) Regulation in Metabolism. Interscience, Wiley, New York.

Odum, H.T. & Pinkerton, R.C. (1955) Time's speed regulator: The optimum efficiency for maximum power output in physical and biological systems. Am. Sci., 43, 331-343.

Ott, J.A. (1981) Adaptive strategies at the ecosystem level: examples from two benthic marine systems. P.S.Z.N.I. Marine Ecology, 2, 113-185.

Pörtner, H.O. et al. (1984) Anaerobiosis and acid-base status in marine invertebrates: a theoretical analysis of proton generation by anaerobic metabolism. J. Comp. Physiol., 155, 1-12.

di Prampero, P.E. (1981) Energetics of muscular exercise. Rev. Physiol. Biochem. Pharmacol., 89, 143-222.

Prigogine, I. (1967) Introduction to Thermodynamics of Irreversible Processes, 3rd Edn. Interscience, Wiley, New York.

Prigogine, I. (1980) From Being to Becoming. Time and Complexity in the Physical Sciences. Freeman, New York.

Putzer, V. (1985) PhD. Thesis, Univ. Innsbruck.

Schrödinger, E. (1944) What is Life? Cambridge Univ. Press, London.

Shick, J.M. et al. (1983) Anoxic metabolic rate in the mussel Mytilus edulis L. estimated by simultaneous direct calorimetry and biochemical analysis. Physiol. Zool., 56, 56-63.

Sidell, B.C. & Beland, K.F. (1979) Lactate dehydrogenases of Atlantic hagfish: physiological and evolutionary implications of a primitive heart isozyme. Science, 207, 769-770.

Stebbing, A.R.D. & Heath, G.W. (1984) Is growth controlled by a hierarchical system? Zool. J. Linn. Soc., 80, 345-367.

Stouthamer, A.H. (1977) Energetic aspects of the growth of microorganisms. In: Microbial Energetics, Eds. B.A. Haddock & W.A. Hamilton, pp. 285-315. Cambridge Univ. Press, London.

Stucki, J.W. (1980) The optimal efficiency and the economic degrees of coupling of oxidative phosphorylation. Eur. J. Biochem., 109, 269-283.

Talsma, W.R. (1980) Wiedergeburt der Klassiker. Anleitung zur Entmechanisierung der Musik. Wort und Welt Verlag, Innsbruck.

Thauer, R.K. et al. (1977) Energy conservation in chemotrophic anaerobic bacteria. Bacteriol. Rev., 41, 100-180.

Westerhof, H.V. et al. (1983) The dynamic efficiency of microbial growth is low, but optimal for maximal growth rate. Proc. Nat. Acad. Sci. USA, 80, 305-309.

Wieser, W. (1985) A new look at energy conversion in ectothermic and endothermic animals. Oecologia, 66, 506-510.

Wijsman, T.C.M. (1976) Adenosine phosphates and energy charge from different tissues of Mytilus edulis L. under aerobic and anaerobic conditions. J. Comp. Physiol., 107, 129-140.

Wilkie, D.R. (1974) The efficiency of muscular contraction. J. Mechanochem. Cell Motility, 2, 257-267.

Yanagida, T. et al. (1985) Sliding distance of actin filament induced by a myosin crossbridge during one ATP hydrolysis cycle. Nature, 316, 366-369.

Zebe, E. et al. (1981) The energy metabolism of the leech Hirudo medicinalis in anoxia and muscular work. J. Exp. Zool., 218, 157-163.
de Zwaan, A. (1983) Carbohydrate catabolism in bivalves. In: The Mollusca, Vol. 1, Ed. P.W. Hochachka, pp. 137-175. Academic Press, New York.

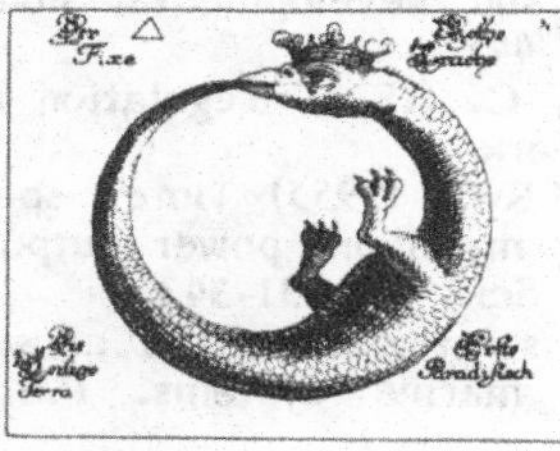

GROWTH AND RESOURCE ALLOCATION

R.M. Sibly
P. Calow

INTRODUCTION

Rules governing resource allocation within organisms are fundamental to an understanding of growth patterns. Thus growth rates and hence growth curves (size versus time/age) of whole organisms depend upon the allocation of resources between anabolic (production) and catabolic (non-production) processes and/or between fast- and slow-growing tissues and parts. Similarly the relative growth of tissues/organisms/parts and hence the shape of organisms, depends upon the relative allocation of resources between these components. It can readily be appreciated, therefore, that the principles of resource allocation behind both the growth of whole organisms and their parts are similar and so there ought to be a general theory of growth. Our aims, here, are to develop such a theory on the assumption that growth patterns evolve according to neoDarwinian principles. Thus we try to find functional explanations of growth patterns - animals with a particular growth pattern (assumed to be heritable) develop faster, survive better or breed more than those without. In the first section we show how this approach can be applied to whole-organism growth patterns and then, in the rest of the chapter, we concentrate on the relative growth of parts. The reason for this emphasis is that whole-organism growth can be considered as the sum of growth patterns of parts. Moreover, the functional basis of whole-organism growth has been considered in detail elsewhere (Sibly et al., 1985).

WHOLE-ORGANISM GROWTH

Any gene promoting growth (i.e. faster development to reproductive maturity) will be favoured by natural selection unless it is associated with a fitness cost of some kind (see below). This statement holds in all populations with plausible rates of increase or decline except when better conditions for offspring survival and development are made available by delaying reproduction (Sibly & Calow, submitted). In other words there is always selection for faster growth if it incurs no fitness costs. In this case we expect growth to be as fast as possible, pushing up against some physiological or nutritional constraint. The alternative possibility is that rapid growth incurs fitness costs, and in this case it is to be expected that faster growth will have been favoured up to the point that the fitness benefits associated with it just outweigh costs.

These two possibilities provide the basis for all the current theories attempting to explain the sigmoid form of growth curves that applies widely throughout the Animal Kingdom (example in Fig. 1). In theories of the first sort the decline in growth rate as the animal approaches maturity is held to be due to a change in the developmental constraint. As an example of this sort of theory, consider the suggestion by Huxley (1932) that there are two incompatible kinds of growth, one (usually occurring first) involving specialist differentiation of tissue, and the other involving merely the enlargement of tissue whose type is already established. As he put it: "In the first, the general form of the part is being laid down, and this process is accompanied by very rapid alterations of form, and by marked histological changes; in the second, histological changes are absent or of an entirely secondary nature, and the form changes are confined to quantitative alterations in the proportions of the definitive structural plan." Since <u>ex hypothesi</u> these two kinds of growth cannot occur simultaneously, they impose a constraint on growth patterns in general. An influential development of this model by Weiss & Kavanaugh (1957) distinguishes between generative mass (capable of growth) and differentiated mass (incapable of growth), and here sigmoid growth arises from the way controlling factors cause the former to be transformed to the latter.

A related though distinct theory due to Ricklefs (1979) holds that there are two incompatible processes, these being growth and (mature) function. It is supposed that mature function only becomes possible when cells are differentiated. Thus as the animal matures, more and more of the embryonic cells differentiate and start to function. As a consequence there are fewer embryonic cells left, and so growth rate declines. This therefore provides one explanation of sigmoid form in growth curves.

In the second kind of theory the decline in growth rate as the animal approaches maturity is held to be an evolved optimal strategy. The declining growth rate is hypothesized to be an optimal balance between the fitness gains of growing faster and supposed fitness costs (e.g. increased

Fig. 1. An example of a sigmoid growth curve in the Barn Owl <u>Tyto alba</u> (after Ricklefs, 1968 and O'Connor, 1984).

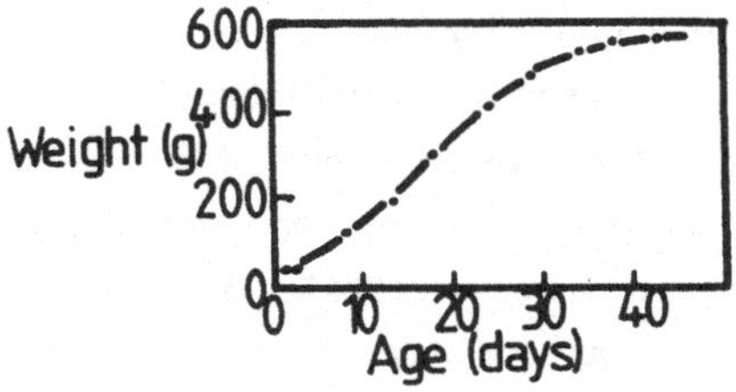

Fig. 2. Hypothetical illustration of how three populations may not show a positive mortality-growth association (broken line) even if each adopts the optimal strategy (*) seen in the context of its own mortality-growth constraint curve (solid lines).

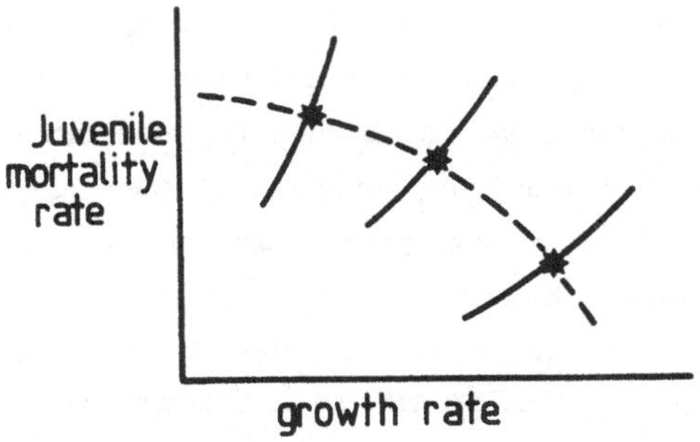

mortality) thereby incurred. For example Lack (1968) suggested that growth rates might be limited by the need to form fat reserves as insurance against temporary food shortage. These fat reserves are a means of avoiding starvation - and could be sacrificed for faster growth - but only by increasing the risk of death from starvation. This provides one of the classic examples of how growing faster can lead to a mortality cost.

Attempts to identify the dependence of mortality on growth rate have relied on comparative studies of the performance of related species, and have shown a link between these variables in, for example, echinoderms (Ebert, 1985), but not in birds (O'Connor, 1984). However interspecific comparison provides a somewhat dubious means of investigation since mortality-growth dependencies may vary between species, so that even if all individuals pursue optimal strategies with respect to their own mortality-growth curve, no relationship appears in comparative studies (example in Fig. 2).

Although it has been profitable to contrast these two sorts of theory, closer examination suggests some of the differences may be more apparent than real. For example Ricklefs' theory of an incompatibility of growth and function may be more properly thought of as an example of the second type of theory rather than of the first, according to the following argument. The function of an organ (e.g. leg muscle) can only be assessed by examining the uses to which the organ has been put in the circumstances in which the ancestors of present individuals evolved. In the case of juvenile walking animals, major functions of leg muscles are escape from predators and also foraging. The former obviously reduces mortality risk and so by investing in leg muscles the mortality risk is lowered. Function was provided, in Ricklefs' theory, at the expense of growth, so it follows that growth rates could be increased at a cost in terms of mortality. Ricklefs' theory, at first sight an example of the first type of theory, turns out on closer examination to be an example of the second.

SIZE CHANGE OF PARTS

The description of the changing differential growth of different organs and structures with age (example in Fig. 3) was one of the major achievements of biology in the interwar years, due in particular to J.S. Huxley. He suggested (Huxley, 1932) that during growth the changing sizes of any two organs, x_i and x_j, can be described by a relationship of the form

$$x_j = Bx_i^{\alpha} \qquad (1)$$

Fig. 3. Increasing weight of brain and liver plotted against increasing weight of heart in man. Data from Thompson (1942).

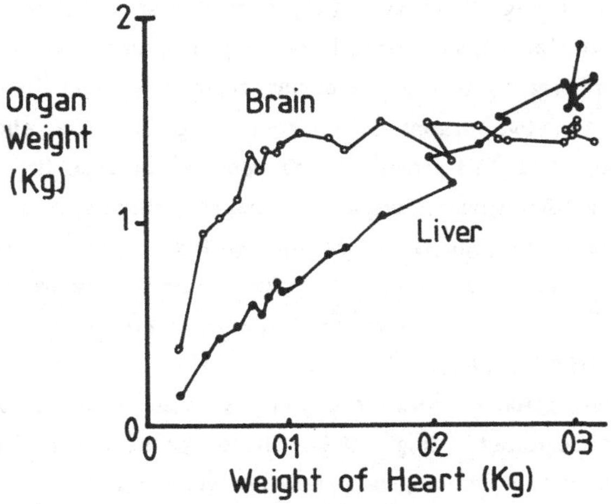

Fig. 4. The three types of growth curve referred to in the text.

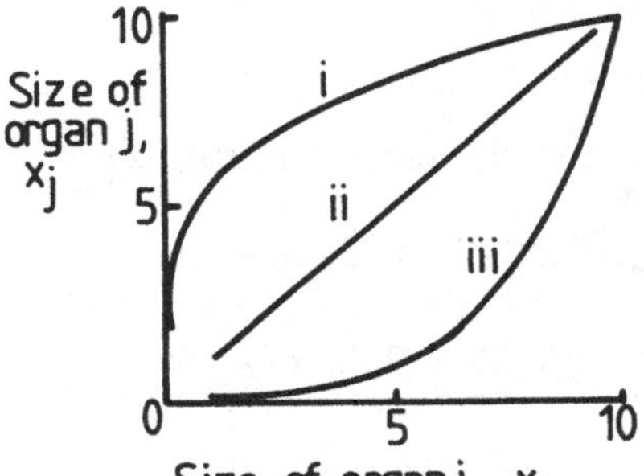

where B = constant and α is often referred to as an allometric coefficient. Huxley classified growth patterns as follows:

(i) $\alpha < 1$

(ii) $\alpha = 1$

(iii) $\alpha > 1$

These are illustrated in Fig. 4. In type (i) growth curves, relative to organ i organ j grows fast initially and slower later, e.g. because the allocation of resources shifts from organ j to organ i during growth. In type (ii) growth the relative growth of the two organs is constant, e.g. because allocation of resources is constant. Type (iii) growth is the opposite of type (i).

Relative to the heart, organs which grow fastest initially (type i) in man include kidney, liver, lungs and brain (see Fig. 3). Organs which grow at the same rate as the heart (type ii) include spleen. Organs which grow slowly initially but fast later relative to the heart (type iii) include the reproductive organs (Brody, 1945).

The description of relative growth has been elaborated since the war, notably by von Bertalanffy (e.g. 1960). There has been some attempt to develop functional models explaining why relative growth follows the observed patterns (e.g. Bryden, 1969) - but as in other branches of biology these have proved difficult to frame rigorously and to document and test experimentally.

Fig. 5. Allocation of resources between different organs. At each time t a fraction u_1 (t) of production P(t) goes to body musculature which is then of size x_1 (t), a fraction u_2 (t) goes to organ No. 2, which is then of size x_2 (t), and so on.

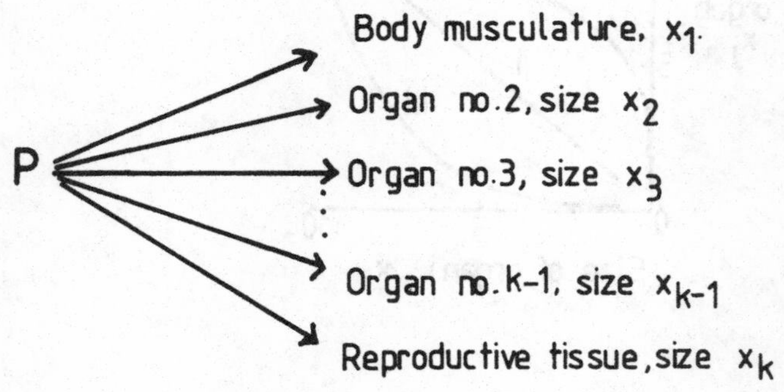

FUNCTIONAL MODEL

During growth, the resources available for the production of new tissue are distributed differentially between the different organs of the body (Figs. 3 & 4). At each instant of time, therefore, resources will be allocated in a well-defined fashion, as shown in Fig. 5. However as the animal grows the allocation may change. We should therefore think of growth as a dynamic strategy of allocation, changing in time. Referring to Fig. 5 we see that the organs are labelled 1, 2, ... k, with organ No. 1 being body musculature, and organ No. k representing reproductive tissue. A fraction u_1 (t) of resources is allocated to body musculature at time t, a fraction u_2 (t) to organ 2, u_3 (t) to organ 3 and so on. Thus

$$\sum_{i=1}^{k} u_i(t) \leqslant 1, \text{ and } 1 \geqslant u_i \geqslant 0 \text{ for all } i \tag{2}$$

If the allocation strategy as a function of age is completely specified, moment by moment, from birth until death, for every organ, then we can describe this mathematically by saying that u_i (t) is completely determined for i = 1 to k, and for t = 0 to ∞ . (There is no implication that the animal will actually live for ever, but the above is taken to imply that the allocation strategy is specified however long the animal lives). The above is usually summarized in mathematical shorthand as follows:

the allocation strategy is $\{u_i(t)\} \quad i=1,\ldots k, \; t = (0,\infty)$

or in abbreviated (vector) form $\{\underline{u}(t)\}_{t=(0,\infty)}$ (3)

Referring again to Fig. 5 we see that a fraction u_i (t) of resources is allocated to organ i at time t, when organ size is x_i. The rate of growth of the organ is dx_i/dt, which we will write $\dot{x}_i$. The total rate of production of new tissue is P, and a fraction u_i consists of the production of new tissue in organ i, in the form of growth of that organ. Hence

$$\dot{x}_i = u_i P \quad i = 1,2\ldots k \tag{4}$$

Since these organs are a functional part of the body, death would occur in most cases if an organ was removed or was absent, and death might also occur if the organ was of an inadequate size. Thus mortality rate, which we shall call μ, must be a mathematical function of organ size and we can plausibly assume it to be a decreasing function (at least over the range of sizes smaller than that found in nature). (We shall return to this later when we shall refer to it as Assumption 2). The same considerations apply to all the body's organs: mortality rate depends on all of them, i.e.

$$\mu = \mu(x_1, x_2, \ldots x_k) \tag{5}$$

$\mu(t)$ is necessarily related to $S(t)$, the probability of an individual surviving to age t, by the equation

$$\frac{dS}{dt} = -\mu S \quad \text{(Sibly et al., 1985)} \tag{6}$$

Organ 1 was taken to be body musculature. Allocations here influence behaviour and this in turn determines the rate of resource acquisition. In order to simplify the mathematics we shall assume that the rate of production (including reproduction) depends only on the amount of body musculature, i.e.

$$P = P(x_1) \tag{7}$$

Similar arguments apply to fecundity (i.e. birth rate), which will be written as n. Thus it will be assumed that fecundity is only influenced by body musculature, i.e.

$$n = n(x_1) \tag{8}$$

We now have a complete specification of the system in terms of growth rates and consequent mortality and fecundity (equations (2) - (8)). These together are sufficient to calculate the neoDarwinian fitness. Following Sibly & Calow (1986) the fitness, F, of a gene in a specified environment is given as the root of

$$1 = \tfrac{1}{2}\int_0^\infty e^{-Ft}\, S(t)\, n(x_1(t))\, dt \tag{9}$$

where $S(t)$ is the probability of individuals carrying the gene surviving to age t, at which age each gives birth to $n(t)dt$ offspring in time interval (t, t+dt). Half of these offspring (on average) receive a copy of any allele present in a parent. As a simplification we assume that both sexes have the same life cycle and individuals mate with others of the same age. The genetic system can be either haploid or diploid. We assume that the age structure is stable.

We are now in a position to calculate the optimal growth strategy, by which is meant the allocation strategy $\{\underline{u}(t)\}_{t=(0,\infty)}$ which maximizes fitness, F (defined by equation (9)) subject to constraints

$$\sum_{i=1}^{k} u_i(t) \leq 1, \quad 0 \leq u_i \leq 1 \text{ for all } i \tag{2}$$

$$\dot{\underline{x}}(t) = \underline{u}(t)\, P(x_1) \tag{4}$$

$$\dot{S} = -\mu S \tag{6}$$

and initial conditions $\underline{x}(0) = \underline{x}_0$, $S(0) = 1$.
It has been shown (Sibly et al., 1985) that life cycles (i.e. $\{\underline{u}(t)\}_{t=(0,\infty)}$ schedules) that maximize F in equation (9) can be found by the following procedure. Find the life cycle that maximizes the function

$$\Phi = \tfrac{1}{2}\int_0^\infty e^{-Rt} S(t)\, n(x_1(t))\, dt \tag{10}$$

with respect to $\{\underline{u}(t)\}_{t=(0,\infty)}$ where R is an arbitrary constant. Do this for a range of different values of R and find the one for which the maximum value of Φ is 1. The number R and the associated allocation pattern are respectively the maximum attainable value of F and the corresponding optimal strategy. Hence we seek an allocation strategy $\{\underline{u}(t)\}_{t=(0,\infty)}$ that maximizes Φ, i.e.

$$\max_{\underline{u}(t)} \Phi = \tfrac{1}{2}\int_0^\infty e^{-Ft} S(t)\, n(x_1(t))\, dt \tag{11}$$

Using Pontryagin's Maximum Principle (Sibly et al., 1985) this is equivalent to

$$\max_{\underline{u}(t)} H = \tfrac{1}{2}e^{-Ft} S(t)\, n(x_1) + \sum_{i=1}^{k} u_i P(x_1)\lambda_i(t) - \mu(x_1, x_2, \ldots x_k)\, S(t)\lambda_{k+1}(t) \tag{12}$$

where $\lambda_1, \lambda_2, \ldots \lambda_{k+1}$ are continuous functions of time to be determined later, specified by

$$0 = \lambda_1(\infty) = \lambda_2(\infty) = \ldots = \lambda_{k+1}(\infty) \tag{13}$$

from Sibly et al., (1985 equation (12)) and

$$\frac{d\lambda_1}{dt} = -\tfrac{1}{2}\frac{dn}{dx_1} e^{-Ft}S - \sum_{i=1}^{k} u_i\lambda_i \frac{dP}{dx_1} + \frac{\partial\mu}{\partial x_1}\lambda_{k+1}S \tag{14}$$

$$\frac{d\lambda_i}{dt} = \frac{\partial\mu}{\partial x_i}\lambda_{k+1}S \quad \text{for } i = 2, 3, \ldots k \tag{15}$$

$$\frac{d\lambda_{k+1}}{dt} = -\tfrac{1}{2} e^{-Ft}n + \mu\lambda_{k+1} \tag{16}$$

from Sibly et al., (1985 equation (11)). Note that following Alexander (1982) and Sibly et al., (1985) S is formally considered to be a 'state variable'.

Equation (16) can be solved using boundary condition (13) to give

$$\lambda_{k+1} = \frac{1}{2S(t)} \int_t^\infty e^{-F\tau} S(\tau)\, n\, d\tau \tag{17}$$

(Sibly et al., 1985). It is interesting to note that $\lambda_{k+1}(t)$ is equal to Fisher's reproductive value (Fisher, 1930) multiplied by e^{-Ft}. Given the value of $\lambda_{k+1}(t)$

it is possible to obtain a formal solution of equation (15) using boundary condition (13) as

$$\lambda_i(t) = -\int_t^\infty \frac{\partial\mu}{\partial x_i} \lambda_{k+1} S d\tau \tag{18}$$

for i = 2, 3, ... k. Since λ_{k+1} S is necessarily positive and $\frac{\partial\mu}{\partial x_i}$ can be assumed negative it follows that λ_i (t) is necessarily positive. Furthermore for any j between 2 and k

$$\lambda_j(t) = -\int_t^\infty \frac{\partial\mu}{\partial x_j} \lambda_{k+1} S d\tau \tag{19}$$

Thus λ_i and λ_j are very closely related, via the functions $\frac{\partial\mu}{\partial x_i}$ and $\frac{\partial\mu}{\partial x_j}$

The optimal pattern of growth can be characterised using equation (12) to plot contours of equal H on the trade-off curve between u_i and u_j as in Fig. 6. It follows from equation (12) that these contours are straight lines, as shown in Fig. 6, increasing in H away from the origin. We can now identify the following cases:

a. If the gradient of the H-lines is steeper than -1 then the optimal strategy is $u_i = u_{i\ max}$, $u_j = 0$.

b. If the slope of the H-lines is shallower than -1 then the optimal strategy is $u_j = u_{j\ max}$, $u_i = 0$.

c. If the lines of equal H are parallel to the trade-off line then some intermediate values ($0 \leq u_i \leq 1, 0 \leq u_j \leq 1$) are optimal.

Cases a. and b. imply bang-bang growth - all resources go first to one organ and then to the other. Only case c. can imply simultaneous growth of the two organs. Thus the allometric equation can only follow from case c. Formally the slope of the equal-H lines ($-\lambda_i/\lambda_j$) is then equal to the slope of the trade-off line (-1), so

$$\lambda_i(t) = \lambda_j(t) \quad \text{for all } t>0 \tag{20}$$

Hence from equations (18) and (19)

$$\int_t^\infty \frac{\partial\mu}{\partial x_i} \lambda_{k+1} S d\tau = \int_t^\infty \frac{\partial\mu}{\partial x_j} \lambda_{k+1} S d\tau \tag{21}$$

and therefore $\frac{\partial\mu}{\partial x_i} = \frac{\partial\mu}{\partial x_j}$ (22)

throughout life.

This formula can be used to calculate the optimal organ sizes from knowledge of the selection pressure μ(x). For example if

$$\mu = \frac{1}{x_j} + \frac{1}{B^2 (2\alpha-1) x_i^{2\alpha-1}} \qquad (23)$$

then $\frac{\partial \mu}{\partial x_j} = -\frac{1}{x_j^2}$ and $\frac{\partial \mu}{\partial x_i} = -\frac{1}{B^2 x_i^{2\alpha}}$

and therefore by equation (22)

$$-\frac{1}{x_j^2} = -\frac{1}{B^2 x_i^{2\alpha}}$$

which can be rearranged to give

$$x_j = B x_i^{\alpha}$$

which is the allometric equation as in equation (1).

INVERSE OPTIMALITY

Assuming that the optimization argument elaborated in the last section is correct, then it is possible to consider what inferences can be made about mortality dependencies (or the contribution of different organs to the overall survivorship of the organism that carries them) from observations of relative growth. These are:

Fig. 6. The trade-off between u_i and u_j (solid line) is a straight line with slope - 1, if the other u s are fixed. The trade-off is given by equation (2), which reduces to $u_i + u_j$ = constant, for some constant between 0 and 1. Hypothetical contours of equal H are superimposed as broken lines (contours of equal H are defined by equation (12)). H increases away from the origin. * denotes the optimal strategy. (a), (b) and (c) correspond to the three cases discussed in the text.

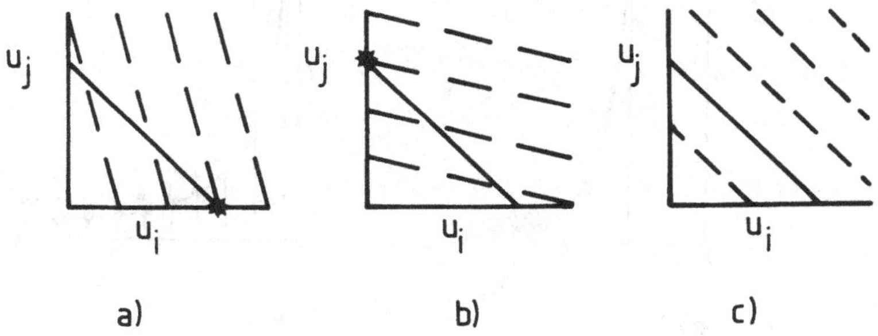

Fig. 7. (a) and (b). Possible relationships between mortality rate and organ size (see text). The optimal strategy is for organ i to grow faster than organ j early in life but for organ j to grow faster later, as shown in (c) and (d) (numbers refer to days since birth). The equations of the curves are $\mu(x_i) = x_i^{-4}$ (a and c); $\mu(x_j) = x_j^{-1}$ (b and d).

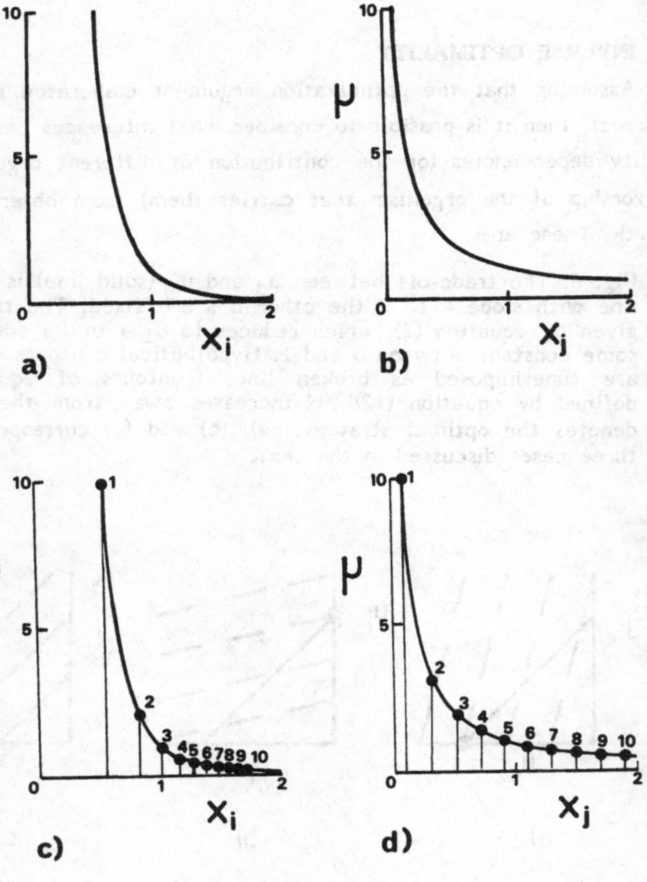

1. The size relationships between most organs and between individual organs and the size of the whole organism usually conform to the allometric equation. Hence this implies conformity to case c (see previous section), so it follows that the allocation of resources between organs is such that unit size change of each makes the same contribution to reducing mortality (increasing survivorship) (i.e. equation (22)). This can be taken as a rule of relative growth.

2. Though organs grow simultaneously, their growth rates can differ. This manifests itself as different allometric exponents (α). What can we infer about mortality curves from a knowledge of the allometric exponent? It can be seen intuitively that the organs which obtain a large share of resources are likely to be those that contribute most to future survival. Remember that mortality depends on organ size - presumably if an organ is too small it cannot adequately fulfil its function, and the problem would get increasingly worse if the organ was progressively reduced in size. Hence the curve relating mortality μ to organ size x is convex seen from below, as in Fig. 7. Different organs will have mortality curves of differing curvature. Intuition suggests that early in development resources should go to organs with initially steep, very convex mortality curves (e.g. Fig. 7a) and only subsequently to others (e.g. Fig. 7b). Suppose for concreteness that the mortality curves are inverse power functions as in Fig. 7. Now equation (19) gives the priority weightings of organs 2 ... k as

$$\lambda_i = - \int_0^\infty \frac{\partial \mu}{\partial x_i} \lambda_{k+1} S d\tau$$

Since organs actually grow simultaneously, though at different rates, we know that all the λ_i s are equal at each age, from which it follows that the $\frac{\partial \mu}{\partial x_i}$ s are equal at each age. $\frac{\partial \mu}{\partial x_i}$ is the reduction in mortality that results from unit increase in organ size, so it follows that organs grow in such a way that each makes the same contribution to reducing mortality. This is illustrated in Fig. 7c and d. Note that at each age the slope of $\mu(x_i)$ equals the slope of $\mu(x_j)$. Supposing, as seems reasonable, the mortalities associated with the two organs are independent of each other, we can write

$$\mu = \frac{A}{x_i^a} + \frac{B}{x_j^b} \tag{24}$$

for some constants A, B, a and b to be determined. The slope of $\mu(x_i)$ is $-aAx_i^{-a-1}$ and the slope of $\mu(x_j)$ is $-bBx_j^{-b-1}$ and as shown above these must be equal at each age. Therefore

$$-aAx_i^{-a-1} = -bBx_j^{-b-1}$$

$$x_j = \left(\frac{bB}{aA}\right)^{\frac{1}{b+1}} x_i^{\frac{a+1}{b+1}}$$

Hence the allometric exponent, α , equals $\frac{a+1}{b+1}$. It follows that if $\alpha > 1$ then $a > b$, so the mortality curve for organ i is more convex and initially steeper than that for j, which is why it gets a greater share of resources early in life. The converse holds if $\alpha < 1$.

In summary, organs which develop fast early in life are predicted to have more convex and initially steeper mortality curves (Fig. 7a) than those developing late (Fig. 7b).

DISCUSSION

The model shows us that, subject to a number of plausible assumptions, the optimal growth strategy can always be calculated, and obeys equation (22). Furthermore equation (18) tells us that an organ for which $\frac{\partial \mu}{\partial x_i}$ is bigger (i.e. its size has more effect on survivorship) will have bigger λ_i - and so will receive greater priority under the optimal strategy. So resources are allocated between organs on the basis of their ability to reduce mortality rate (improve survivorship). By making the specific assumption that mortality curves are inverse power functions as given by equation (24) it is possible to make deductions about the shapes of the mortality curves given that observed growth patterns are optimal and these were discussed in the last section.

An experimental programme, based on this analysis, could involve:

(a) Comparing αs in animals subject to different levels of organ-dependent risks. For example, because the liver is a detoxifying organ, one might find faster liver growth in animals suffering higher concentrations of a toxin. This would imply developmental plasticity, and if this had been optimized by selection, the liver growth would increase with toxin concentration. On the other hand even if there was no developmental plasticity, liver growth might still vary between populations that had undergone selection at different levels of pollution. Another example is provided by gut size, which increases with body size if the diet deteriorates, at least in some birds (Sibly, 1981).

(b) Altering organ size experimentally, e.g. by hormone treatment or by exploiting developmental plasticity (cf. Bryant & Simpson, 1984;

Buul-Offers, 1984), and assessing organ-dependent mortality in a variety of test environments. This approach should yield the $\mu(x_i)$ curves directly. For example, birds with guts of different sizes could be obtained by maintaining them on diets of differing quality. If challenged with a poor diet, birds with small guts might be unable to cope, and so risk starvation. This experiment would only work if the animals were kept in the test environment for a much shorter period than is necessary for gut size to acclimatize (about 25 days).

Finally we should consider to what extent the assumptions that are written into the model are reasonable. Unfortunately there are a large number of these, and it would be tedious, at this stage, to examine them all closely. However the major ones are as follows:

Assumption 1, that resources are divided as in Fig. 5, so that $\sum_{i=1}^{k} \dot{x}_i \leq P$ and $\sum_{i=1}^{k} u_i \leq 1$. This involves some simplifications but seems to us a reasonable first step. Since we later went on to consider all possible allocation strategies $\{\underline{u}(t)\}_{t=(0,\infty)}$ we are in effect assuming that the options open to an animal, in evolutionary terms, are to vary the distribution of resources between tissues, i.e. to change the u_i variables in Fig. 5. The allocation is not constant in time, and different changes are possible at different times. These options define the 'phenotypic options set' within which the optimal strategy is calculated.

Assumption 2, that mortality is a decreasing function of the sizes of the organs, at least over the range of sizes smaller than that found in nature. This seems thoroughly reasonable to us, (although it would be interesting to investigate a U-shaped function, but this has not so far been attempted) and provides a link with theories of whole-organism growth discussed in an earlier section of this chapter that assume that faster growth incurs mortality costs. Suppose, for example, that growth were assessed on the basis of overall body length (as a measure of 'body musculature'). It follows from Assumption 1 that the resources allocated to body musculature can be varied at the expense of other organs. To keep things simple, focus on just one other organ - fat reserves. Thus growth rate of body musculature can be increased at the expense of fat reserves. But if starvation is a possibility this may imply an increase in mortality rate. Therefore faster growth rate can incur a mortality cost, as assumed by some of the theories of whole-organism growth that we have already discussed.

In nature the dependence of mortality rate on organ sizes need not be a deterministic one, and further work is needed to investigate the effect on predictions of stochastic variation in this relationship. Other simplifying assumptions were made in equations (7) and (8), with the objective of keeping the mathematics relatively simple.

We are extremely grateful to Dr. N.K. Nichols for commenting on the mathematics, and to Professor's D. Ware and R. Dunbrack for reviewing the manuscript.

REFERENCES

Alexander, R.M. (1982) Optima for Animals. Edward Arnold, London.

Bertalanffy, L. von (1960) Principles and theory of growth. In: Fundamental Aspects of Normal and Malignant Growth, Ed. W.W. Nowinski. pp. 137-259. Elsevier, Amsterdam.

Brody, S. (1945) Bioenergetics and Growth. Hafner Publishing Co. New York.

Bryant, P.J. & Simpson, P. (1984) Intrinsic and extrinsic control of growth in developing organs. Q. R. Biol., 59, 387-415.

Bryden, M.M. (1969) Regulation of relative growth by functional demand: its importance in animal production. Growth, 33, 143-156.

Buul-Offers, S. van. (1984) The effects of pituitary, thyroid, pancreatic and sexual hormones on body length and weight and organ weights of snell dwarf mice. Growth, 48, 101-119.

Ebert, T.A. (1985) Sensitivity of fitness to macroparameter changes: an analysis of survivorship and individual growth in sea urchin life histories. Oecologia, 65, 461-67.

Fisher, R.A. (1930) The Genetic Theory of Natural Selection. Clarendon Press, Oxford.

Huxley, J.S. (1932) Problems of Relative Growth. Methuen, London.

Lack, D. (1968) Ecological Adaptations for Breeding in Birds. Methuen, London.

O'Connor, R.J. (1984) The Growth and Development of Birds. John Wiley, Chichester.

Ricklefs, R.E. (1968) Patterns of growth in birds. Ibis, 110, 419-451.

Ricklefs, R.E. (1979) Energetics, constraint and compromise in avian postnatal development. Biol. Rev., 54, 269-290.

Sibly, R.M. (1981) Strategies of digestion and defecation. In: Physiological Ecology: an Evolutionary Approach to Resource Use. (Eds. C.R. Townsend & P. Calow). pp. 109-139. Blackwell Scientific Publications, Oxford.

Sibly, R.M. & Calow, P. (1986) Physiological Ecology of Animals: an Evolutionary Approach. Blackwell Scientific Publications, Oxford.

Sibly, R.M., Calow, P. & Nichols, N. (1985) Are patterns of growth adaptive? J. Theor. Biol., 112, 553-74.

Thompson, D'A. W. (1942) On Growth and Form. Cambridge University Press, Cambridge.

Weiss, P. & Kavanau, J.L. (1957) A model of growth and control in mathematical terms. J. Gen. Physiol., 41, 1-47.

MAINTENANCE AND REPAIR PROCESSES IN RELATION TO SENESCENCE: ADAPTIVE STRATEGIES OF NEGLECT

T.B.L. Kirkwood

INTRODUCTION

Repair serves to prolong life; senescence brings it to a close. To what extent are these two processes biologically related? Under what patterns of natural selection have they evolved?

The idea that senescence may be linked with the failure of life-maintenance mechanisms has a long history (for reviews, see Comfort, 1979; Kirkwood & Cremer, 1982). The outward signs of senescence are certainly consistent with an accumulation of wear and tear. Skin wrinkles, joints stiffen, senses weaken, and so on. These are all features which we can identify quite readily with the progressive accumulation of defects in objects we manufacture, such as computers or cars. There are, however, good reasons why we should hesitate before accepting that wear and tear will inevitably bring about death. Firstly, senescence is not intrinsic to all living organisms. Bacteria and many eukaryotic microorganisms can divide indefinitely, many plants are capable of unlimited vegetative growth, and some simple animals, such as coelenterates, have regenerative powers which appear to allow their indefinite survival (Comfort, 1979). Secondly, as Williams (1957) observed, higher organisms invest so much effort in morphogenesis and maturation that it would really be quite surprising if they were fundamentally incapable of merely maintaining what is already formed.

In this paper, the factors governing evolution of senescence and repair are reviewed and the theory is advanced that senescence has evolved through optimizing the allocation of metabolic resources so that only limited effort is put into somatic maintenance and repair (Kirkwood, 1977, 1981; Kirkwood & Holliday, 1979, 1986). The implications of the theory for evolution of life histories are discussed, together with ways to test the theory experimentally.

SELECTION FOR MAINTENANCE AND REPAIR

One of the striking features of living organisms is their ability to repair themselves. Repair is not a fundamental biological necessity but would have been a vital adaptation if organisms were to progress beyond the simple self-replicating polymers which presumably were the earliest forms of life. Today there is probably no species which is not capable of some form of repair.

In some instances, repair is highly visible, as when an entire anatomical structure is lost and regenerated, or when lacerated tissue heals together and continuity of the skin is restored. At the other extreme is a continual, but inconspicuous background of minor repair activities. The latter include the excision and correction of DNA lesions, the repair of damage to cell membranes from the destructive effects of highly reactive metabolic wastes, such as free radicals, the mopping-up of abnormal proteins and polynucleotides by hydrolytic scavenging enzymes, and the relentless proof reading of macromolecule synthesis to weed out errors (see Kirkwood et al., 1986). In addition to these processes of intracellular repair, there is in multicellular organisms the continual mitotic replacement of cells which die.

The two ends of this spectrum of repair activities can be termed 'emergency' repair and 'maintenance' repair, respectively. The principles governing evolution of either type of repair function were described in detail elsewhere (Kirkwood, 1981) and will be reviewed only briefly here. For any repair function to evolve, three basic conditions must be fulfilled. Firstly, organisms must be able to survive the damage at least for long enough for repair to take place. Secondly, the information to restore the damaged part to its undamaged form must be available. Thirdly, the overall benefit of repair, in terms of its effect on reproductive fitness, must outweigh the overall cost.

For severe forms of damage the fulfilment of all three conditions is less likely than for minor damage. In consequence, the phylogenetic distribution of emergency repair functions is expected to be less uniform than that of maintenance repair. This is presumably the reason why regeneration ability varies markedly between species (Kirkwood, 1981; Reichman, 1984). Maintenance repair, on the other hand, is expected to be much more evenly distributed, and similar mechanisms are observed in virtually all species.

In relation to senescence our main interest centres on maintenance repair. This is because, although emergency repair also prolongs life, it is by maintenance repair that the gradual and insidious accumulation of minor defects can be slowed. Even if a particular maintenance repair function is available, it need not be used to full capacity. Before exploring the relationship between senescence and maintenance repair more closely, however, we consider first how senescence may be defined and by what general process of natural selection it may have evolved.

NATURE AND ORIGINS OF SENESCENCE

For comparative purposes, senescence is best defined as the sum of those effects which render individuals as they grow older more susceptible to the various factors, intrinsic or extrinsic, which may cause death (Maynard Smith, 1962). Senescence is thus manifest in a population when there is a progressive increase in the age-specific mortality rate, and not otherwise. This actuarial definition is applicable in quite different phyla where the specific features of the ageing process may be quite distinct. The definition does not apply, however, to species which exhibit once only, or semelparous, reproduction (see Kirkwood, 1985). Much confusion has arisen in the gerontological literature over species, such as the Pacific salmon, which dies soon after first spawning. This behaviour has in fact quite different biological significance from the gradual decline in vitality of an organism which reproduces repeatedly during its lifetime and whose life history could, potentially, extend indefinitely. Detailed comparison of these different life-history plans is outside the scope of this paper (see Kirkwood, 1981, 1985), and I deal here only with repeatedly reproducing, or iteroparous, organisms.

Among general theories to account for the evolution of senescence two principal categories can be distinguished. In the first, senescence is seen as a beneficial trait in its own right. These are the 'adaptive' theories. The second category holds that senescence is detrimental, or at best neutral, so that its evolution might be explained indirectly. These are the 'non-adaptive' theories.

Explicit formulations of the adaptive theories are rare, despite the fact that this category enjoys widespread popularity. For this reason, the obvious theoretical weakness of the adaptive view has often escaped notice. The formidable difficulty confronting any adaptive theory of senescence may be seen clearly if one examines its effect on the Malthusian parameter, or intrinsic rate of natural increase, which is defined as the unique real root, r, of the equation:

$$\int_0^\infty e^{-rx}\, l(x)\, m(x)\, dx = 1 \qquad (1)$$

where $l(x)$ is survivorship to age x and $m(x)$ is reproductive rate at age x (e.g. see Charlesworth, 1980). Since a genotype in which senescence occurs will, other things being equal, have lower values of the product $l(x)m(x)$, and hence of r, than a genotype in which senescence does not occur, the senescent genotype will be at a direct selection disadvantage. In fact, proponents of the adaptive view cite, for the most part, benefit at the group or species level as the source of the selection advantage for senescence, the ideas most commonly expressed being that a definite limit to survival prevents over-crowding and promotes genetic change (e.g. Wynne-Edwards, 1962; Woolhouse, 1967). This requires, however, that the selection advantage at the group level should be strong enough to outweigh the disadvantage at the individual level. It is now generally accepted that the circumstances in which this could be true will be exceedingly rare (Maynard Smith, 1976).

If senescence is not directly adaptive, and if one is not to accept that senescence is the inescapable price of biological complexity, two alternatives remain. The first is to suggest that natural selection is simply unable to prevent the deterioration of older organisms because its force becomes too attenuated with age (Haldane, 1941; Medawar, 1952). The second is to suggest that senescence is a by-product of selection for other traits (Williams, 1957). In the latter case, although senescence arises through adaptation it is not in itself advantageous and indeed will remain a negative component of the life history. It is important, therefore, not to confuse Williams' non-adaptive view with the adaptive theories (see also Appendix 1 of Kirkwood & Cremer, 1982).

Central to understanding the origin of senescence is the fact that for an organism which reproduces more than once in its lifetime, events which occur late in the lifespan, whether good or bad, will be of lesser significance than events which occur early. The reason for this is that, regardless of whether or not the mortality rate increases with age, the proportion of survivors becomes steadily smaller and the remaining fraction of their lifetime expectation of reproduction less. The result is that genes with late-acting effects on survivorship and fecundity will be subject to much weaker selection than genes which have their effects early in life (Haldane, 1941; Medawar, 1952; Williams, 1957; Hamilton, 1966; Charlesworth, 1980).

The possibility that this attenuation with age in the force of natural selection is responsible in a very simple way for the evolution of senescence was considered by Medawar (1952). He suggested that selection acting on genes with age-specific times of expression would tend to defer the age of expression of harmful genes, so as to minimize their potential for deleterious effects. Once a harmful gene had been so far deferred that it was expressed at an age when survivorship in the natural environment was effectively zero, natural selection could no longer further postpone its expression, nor could it act to eliminate the gene altogether. Over many generations, there might thus accumulate a heterogeneous collection of late-acting deleterious genes which in the normal environment would not have opportunity to be expressed, but which in a protected environment would combine to handicap severely any individual which by virtue of reduced environmental mortality lived long enough to encounter their detrimental effects. This is very much the pattern that is seen for senescence - clear senility being rare in the wild and common only in captivity - and Medawar suggested such a mechanism could account both for the evolutionary origin and for the present-day nature of the ageing process.

That a simple accumulation of late-acting deleterious genes is a sufficient explanation for the evolution of ageing has been questioned, however, in more recent theoretical and experimental studies (Kirkwood, 1977; Kirkwood & Holliday, 1979; Rose & Charlesworth, 1980). A more plausible view is that senescence is the late deleterious by-product of processes which are beneficial to the organism during the earlier and biologically more important stages of its lifespan (Williams, 1957; Charlesworth, 1980). I turn now to consider a theory derived from an independent line of reasoning but which is compatible with Williams' (1957) general view and can, in a sense, be considered a special instance of it.

EVOLUTION OF SENESCENCE THROUGH OPTIMIZING THE INVESTMENT IN SOMATIC MAINTENANCE AND REPAIR

At the most fundamental level of biological thinking, an organism may be regarded as an entity which takes up energy, primarily in the form of nutrients, from its environment, and ultimately converts this energy into progeny. The law of natural selection asserts that those organisms (strictly, the genes which determine the phenotypes of the organisms) which are most efficient in this process are the ones most likely to survive (see Townsend & Calow, 1981). Of the energy an organism takes in, however, only a fraction is allocated directly to reproduction, the rest being divided among activities such as growth, foraging, and defense, as well as, in particular, the maintenance and repair of the body, or 'soma'. The greater the fraction of energy allocated to one particular activity, the less is available for the others.

The trade-off between allocation of energy to reproduction and allocation of energy to somatic maintenance and repair is crucial to optimizing the life history (Calow, 1977, 1979; Kirkwood, 1977, 1981). Too little investment in somatic maintenance and repair and the organism may die before it can effectively reproduce; too great an investment and the organism will mature late and reproduce only slowly. The question is: does there exist a level of investment in somatic maintenance and repair which is optimal, and if so, is this more or less than the minimum level required to keep the organism in a physiological steady-state by repairing damage as fast as it arises? That such a level exists, at least in theory, was implied by the assertion made earlier that there is no fundamental biological reason why a complex organism should be unable to maintain its physiological integrity indefinitely.

To answer these questions requires that the effects of varying the level of investment in somatic maintenance and repair on survivorship and fecundity as functions of age can be determined. Once these are known, the net effect on the intrinsic rate of natural increase, r, can be determined from equation (1). The optimum level of investment in somatic maintenance and repair will be that which maximizes r, subject to appropriate constraints.

At present, experimental data are not yet available to permit these trade-offs to be evaluated for real organisms. Until this situation changes, some progress can nevertheless be made with plausible models. In the remainder of this paper a simple model will be described which illustrates the main conclusions (Kirkwood & Holliday, 1986). A more general treatment will be given elsewhere (Kirkwood, in preparation).

A convenient representation of the survivorship function, l(x), can be derived from the observation that at least among mammals adult mortality is well described by the Gompertz-Makeham equation

$$\mu_x = \mu_o e^{\beta x} + \gamma ,$$

where μ_x is the age-specific mortality rate at age x and the parameters μ_o, β and γ represent 'basal vulnerability', 'actuarial ageing rate' and age-independent environmental mortality, respectively (see Sacher, 1978; Kirkwood, 1985). [The possibilities of periodic (e.g. seasonal) effects on mortality and of non-Gompertzian mortality curves (e.g. for species where growth is indeterminate and adult mortality initially declines due to increasing body size) are not considered here.]

If a variable ρ is defined to represent the fractional investment in somatic maintenance and repair, $\rho=0$ corresponding to zero repair and $\rho=1$ corresponding to the maximum which is physiologically possible, then β, and probably also μ_o, will be decreasing functions of ρ. To be consistent with the assertion that it is possible, at least in principle, to prevent damage from accumulating by investing enough in repair, it should be assumed that β reaches zero for some intermediate level of repair $\rho=\rho'$ $(0<\rho'<1)$. μ_o, on the other hand, can be expected to decrease steadily across the entire range from $\rho=0$ to $\rho=1$. Finally, it may be noted that although for computing the intrinsic rate of natural increase it is not necessary to know the pattern of mortality across the pre-adult age range (since m(x) is zero), the overall dependence of juvenile survivorship on ρ must be specified in order that the starting point for the adult survivorship function may be known.

For reproductive rate, m(x), the primary parameters affected by ρ will be age at first reproduction, a, which is likely in general to be increased by raising ρ, and the peak reproductive rate, f, which can be expected to decrease with ρ. For an organism with $\rho<\rho'$, i.e. one with $\beta>0$, accumulation of somatic damage will be likely to take its toll on reproductive rate as age increases and this should also be allowed for in the model.

Typical functions giving the required dependence of μ_o, β, a and f on ρ can be assumed as follows: $\mu_o = \mu_{min}/\rho$, $\beta = \beta_o(\rho'/\rho-1)$ for $\rho<\rho'$, $a = a_o/(1-\rho)$, and $f = f_{max}(1-\rho)$. It is additionally assumed that juvenile mortality can be represented by survival of a fraction ρ of neonates, with the Gompertz-Makeham equation applying thereafter, and that reproduction begins at peak rate f at age a, with for $\beta>0$ an age-related decline at the same Gompertzian rate as survivorship. With these assumptions the survivorship and fecundity functions are

$$l(x) = \rho \exp[-(e^{\beta x}-1)\mu_o/\beta - \gamma x]$$

and

$$m(x) = f \exp[-(e^{\beta x}-e^{\beta a})\mu_o/\beta] \qquad x>a$$

with the dependence of μ_o, β, a and f on ρ as above.

The net effects on l(x) and m(x) of varying ρ are shown in Fig. 1. When these effects are combined together to give the dependence of r on ρ, the result is as in Fig. 2.

The model confirms that, at least for the specific formulation treated above, an optimum level of investment in somatic repair does exist.

Fig. 1. Effects on survivorship, l(x), and reproductive rate, m(x), of varying the investment, ρ, in somatic maintenance and repair. The arrows indicate the direction of increasing ρ. (From Kirkwood & Holliday, 1986).

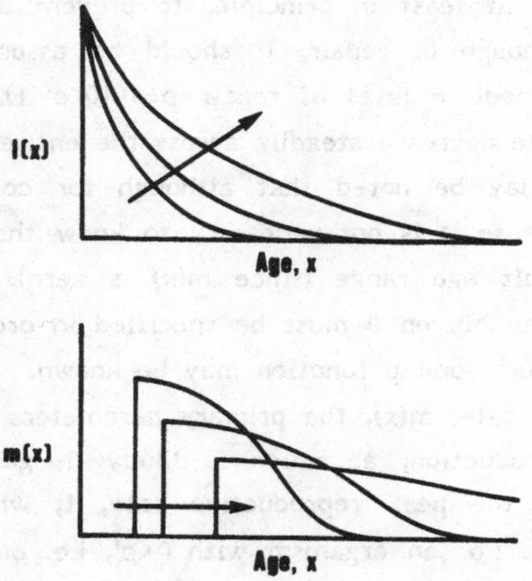

Fig. 2. Relationship between intrinsic rate of natural increase, r, and the level of investment, ρ, in somatic maintenance and repair. For ρ>ρ' it is assumed that the organism is able to repair damage as fast as it arises, so damage does not accumulate. The optimum value ρ=ρ*, i.e. that which maximises r, is found to be less than ρ'. Full details of the model are in the text; parameter values are $\rho'=0.8$, $\mu_{min}=0.005$, $\beta_o=0.5$, $a_o=2$, $f_{max}=10$, $\gamma=0.38$.

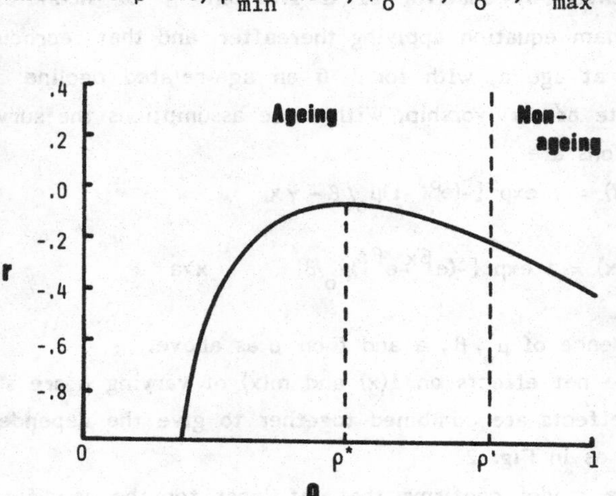

Furthermore, it has been found quite generally that the optimum value of ρ, $\rho=\rho^*$, is <u>less</u> than the minimum, $\rho=\rho'$, above which the actuarial ageing rate is assumed to be zero. This means that even though it may be physiologically possible to set the level of investment in somatic maintenance and repair high enough that damage can be prevented from accumulating, it is not to an iteroparous species' advantage to do so. Instead, it is preferable only to invest sufficient resources in physiological maintenance and repair to ensure that the soma remains viable through its normal expectation of life in the wild, and to use any extra resources liberated by this 'deliberate' neglect of the soma to increase reproduction.

The outcome of the selection process described above will be to bring about the non-adaptive evolution of senescence as a by-product of optimizing the allocation of physiological resources. This view of the evolution of senescence has been termed the 'disposable soma' theory (Kirkwood & Holliday, 1979; Kirkwood, 1981) for its obvious analogy with the manufacture of disposable goods where little is invested in their long-term durability.

DIVERGENCE OF SPECIES' LIFESPANS

For any theory on the evolution of senescence, an important secondary problem is to account for the divergence of species' lifespans. For the disposable soma theory, the question is how the optimum investment in somatic maintenance and longevity may be expected to vary from one ecological niche to another.

The key factor in determining the optimum level of investment in somatic maintenance and repair is the level of environmental mortality to which individuals are exposed. Note that our concern here is not the interaction of the age-specific pattern of mortality with life history, for which extensive theory exists already (e.g. see Stearns 1976, 1977; Sibly & Calow, 1983), but rather with the effect on longevity of varying the overall intensity of environmental mortality within a set life-history plan. If the level of environmental mortality is high, the individual can expect to survive only a short time and there is little point in investing heavily in somatic maintenance and repair. In consequence, when an individual is removed to a protected environment, senescence will occur early and maximum longevity will be short. Conversely, if the level of environmental mortality is reduced, survivorship in the natural environment will be enhanced and there will be advantage to be gained from investing more heavily in somatic maintenance and repair so that the organism does not die of senescence before it has realised its potential for a longer life.

This argument can be formalised within the mathematical framework of the life-history model described above, where environmental mortality is represented through the parameter γ of the Gompertz-Makeham equation. If it is assumed an adaptation leads to a reduction in γ, a compensatory change in one or more of the parameters μ_{min}, β_o, a_o or f_{max} is likely to follow. This may occur either through population density effects, to preserve a long-term stable population size, or as a direct physiological trade-off. For example, evolution of a larger brain may at the same time bring about a reduction both in γ and in the maximal reproductive rate, f_{max}. Introducing these variations into the model, while leaving the other parameters fixed, brings about a resulting change in the optimum value of ρ, with corresponding effects on the lifespan (Table 1).

Table 1
Effect of varying environmental mortality (γ) on the optimum level of investment in somatic maintenance and repair (ρ^*), subject to the constraint that compensatory adjustment in the maximal rate of reproduction (f_{max}) keeps the intrinsic rate of natural increase to be zero at the optimum. (The other parameters of the model are as for Fig. 2). Also shown is the effect on longevity, expressed as the 99th percentile of the lifespan distribution. The time unit is arbitrary, but can be assumed to be years.

γ	f_{max}	ρ^*	Longevity
0.51 (40%)[+]	18.6	.46	8
0.22 (20%)	7.0	.54	12
0.11 (10%)	2.9	.65	28
0.05 (5%)	1.5	.74	57

[+] Figures in brackets show equivalent % mortality per year.

The disposable soma theory therefore predicts a strong positive correlation of the level of environmental mortality with fecundity and a negative correlation with longevity. These correlations are already well recognised from observation of natural populations. The novelty of the disposable soma theory, however, is that it defines a direct mechanism - namely, varying the investment in somatic maintenance and repair - through which they are generated.

SENESCENCE IN HUMAN POPULATIONS

In human populations, certain features of the ageing process contrast markedly with those in most other animal species, and any general theory for evolution of senescence must explain these as well (Kirkwood, 1985; Kirkwood & Holliday, 1986). Firstly, our species is unique in the high frequency with which individuals survive long enough to show clear signs of senescence, especially in the so-called developed countries. Secondly, the human female exhibits a controlled shutdown of reproductive function at menopause, well before maximum specific lifespan is attained. The first of these features suggests a challenge to the idea that selection will usually operate so that senescence remains a latent biological possibility rather than a patent biological fact. The second is often cited by advocates of the adaptive view of ageing as evidence that a strict genetic programme for senescence may exist. In fact, neither of these observations constitutes a serious challenge to the non-adaptive view of senescence, and both may be accounted for in terms of the disposable soma theory.

The high incidence of senescence in modern human societies may be due, in part, to the fact that the rapid advances in social and medical care in the last few centuries will have far outstripped the potential for natural selection to adapt our life history. Nevertheless, senescence is clearly seen, albeit less frequently, in more "primitive" societies and mention of it is found in the earliest human records. The explanation for this may simply be that as environmental mortality was progressively reduced, under the influence of evolving human intellect and the associated trend to living in protected social groups, there came a point where it was no longer selectively advantageous to increase the level of investment in somatic maintenance and repair at the expense of further delaying growth (and with it reproductive maturation) and of reducing reproductive output. If this were the case, as the disposable soma model suggests is plausible (Kirkwood, unpublished results), the continuing selection pressure for further reduction in environmental mortality would have increased the average lifespan while leaving the underlying rate of senescence unchanged. This would have resulted in an increased frequency of survivors into the visibly senescent phase of the lifespan.

As soon as senescence became a significant feature of human or pre-human societies, even if only a small proportion of individuals survived to experience it, long-lived females would have been confronted with the serious risk of attempting to continue reproduction with a senescent soma. Childbearing for humans is in any case made more difficult by the large size of the neonatal brain, and if reproduction were continued throughout the female lifespan, the hazards associated with giving birth would undoubtedly constitute a dominant cause of mortality in older women. It makes sense, therefore, to suppose that the menopause evolved as a means of protecting older women from this risk and of preserving them for the important fitness-enhancing roles of contributing to the successful rearing of their later children and of sharing their accumulated knowledge and experience for the general wellbeing of their kin group. Seen in this way, the menopause is not a primary feature of senescence process, but rather a secondary adaptation to come to terms with it.

TESTING THE THEORY

Despite generations of research, the causes of senescence remain obscure. Comfort (1979, p.17) places blame for this on theory which fails to stimulate incisive experimentation : "Throughout its history the study of ageing ... has been ruinously obscured by theory, and particularly theory of a type which begets no experimental work". For any new theory the challenge is thus well defined.

An attraction of the disposable soma theory is that, on the one hand, it provides a ready explanation for senescence which is directly consistent with the wear and tear nature of the process, while on the other hand, it makes definite predictions which are readily amenable to experimental test. A direct means to test the predictions of the theory, and in the process possibly to illuminate the primary causes of senescence, is provided by comparative study of the efficiency of different systems for somatic maintenance and repair. Over the years, many specific types of damage have been suggested as unique causes of senescence (Sacher, 1980). The disposable soma theory lends support to each of these, but suggests that no single process is likely to account for senescence in all the diverse forms of organism in which it occurs. Those systems for somatic maintenance and repair whose relative efficiencies correlate most strongly with longevity would constitute primary targets for intensive research.

In selecting systems which should be early candidates for

comparative study, some degree of a priori reasoning is likely to pay dividends. In the search for primary causes of senescence, attention should be directed first at the most fundamental level. Among mammals it appears that there may be an intrinsic senescence process limiting the capacity of somatic cell lineages to proliferate (Hayflick, 1977), so there is reason to focus initially on mechanisms for intracellular maintenance and repair. The most fundamental of intracellular maintenance mechanisms are those responsible for the correction or elimination of errors made during the synthesis of macromolecules, and it is suggested that these systems deserve special attention (Kirkwood & Holliday, 1979; Holliday, 1986). Comparative studies on the efficiency of DNA excision repair have revealed already that there are correlations with mammalian longevity (Hart & Setlow, 1974; Francis et al., 1981; Treton & Courtois, 1982; Hall et al., 1984). Should errors in macromolecules prove, however, not to be the primary cause of cellular senescence, investigation would proceed naturally to the higher levels of cellular organisation.

REFERENCES

Calow, P. (1977) Ecology, evolution and energetics: a study in metabolic adaptation. Adv. Ecol. Res., 10, 1-62.

Calow, P. (1979) The cost of reproduction - a physiological approach. Biol. Rev., 54, 23-40.

Charlesworth, B. (1980) Evolution in Age-structured Populations. Cambridge University Press, Cambridge.

Comfort, A. (1979) The Biology of Senescence. 3rd Edition. Churchill Livingstone, Edinburgh.

Francis, A.A. et al. (1981) The relationship of DNA excision repair of ultraviolet-induced lesions to the maximum lifespan of mammals. Mech. Age & Develop., 16, 181-189.

Haldane, J.B.S. (1941) New Paths in Genetics. Allen and Unwin, London.

Hall, K.Y. et al. (1984) Correlation between ultraviolet-induced DNA repair in primary lymphocytes and fibroblasts and species maximum achievable lifespan. Mech. Age. & Develop., 24, 163-173.

Hamilton, W.D. (1966) The moulding of senescence by natural selection. J. Theor. Biol., 12, 12-45.

Hart, R.W. & Setlow, R.B. (1974) Correlation between deoxyribonucleic acid excision repair and lifespan in a number of mammalian species. Proc. Nat. Acad. Sci., USA, 71, 2169-2173.

Hayflick, L. (1977) The cellular basis for biological aging. In: Handbook of the Biology of Aging, Eds. C.E. Finch & L. Hayflick, pp 159-186. Van Nostrand Reinhold, New York.

Holliday, R. (1986) Genes, Proteins and Cellular Ageing. Van Nostrand Reinhold, Philadelphia.

Kirkwood, T.B.L. (1977) Evolution of ageing. Nature, 270, 301-304.

Kirkwood, T.B.L. (1981) Repair and its evolution: survival versus reproduction. In: Physiological Ecology: an Evolutionary Approach to Resource Used, Eds. C.R. Townsend & P. Calow, pp 165-189. Blackwell Scientific Publications, Oxford.

Kirkwood, T.B.L. (1985) Comparative and evolutionary aspects of longevity. In: Handbook of the Biology of Aging, Ed. C.E. Finch & E.L. Schneider, pp. 27-44. Van Nostrand Reinhold, New York.

Kirkwood, T.B.L. & Cremer, T. (1982) Cytogerontology since 1881: a reappraisal of August Weismann and a review of modern progress. Hum. Genet. 60, 101-121.

Kirkwood, T.B.L. & Holliday, R. (1979) The evolution of ageing and longevity. Proc. Roy. Soc. B205, 531-546.

Kirkwood, T.B.L. & Holliday, R. (1986) Ageing as a consequence of natural selection. In: The Biology of Human Ageing, Eds. K.J. Collins & A.H. Bittles, pp. 1-16. Cambridge University Press, Cambridge.

Kirkwood, T.B.L. et al. (1986) Accuracy in Molecular Processes: its Control and Relevance to Living systems. Chapman & Hall, London.

Maynard Smith, J. (1962) Review lectures on senescence. I. The cause of ageing. Proc. Roy. Soc. B157, 115-127.

Maynard Smith, J. (1976) Group Selection. Q. R. Biol., 51, 277-283.

Medawar, P.B. (1952) An Unsolved Problem in Biology. H.K. Lewis, London. (Reprinted in Medawar, P.B. (1981). The Uniqueness of the Individual. Dover, New York).

Reichman, O.J. (1984) Evolution of regeneration capabilities. Am. Nat., 123, 752-763.

Rose, M.R. & Charlesworth, B. (1980) A test of evolutionary theories of senescence. Nature, 287, 141-142.

Sacher, G.A. (1978) Evolution of longevity and survival characteristics in mammals. In: The Genetics of Ageing, Ed. E.L. Schneider, pp. 151-167. Plenum, New York.

Sacher, G.A. (1980) Theory in gerontology. Part I. An. Rev. Gerontol & Geriat., 1, 3-25.

Sibly, R. & Calow, P. (1983) An integrated approach to life-cycle evolution using selective landscapes. J.Theor. Biol., 102, 527-547.

Stearns, S.C. (1976) Life-history tactics: a review of the ideas. Q. R. Biol., 51, 3-47.

Stearns, S.C. (1977) The evolution of life history traits: a critique of the theory and a review of the data. Ann. Rev. Ecol. Syst., 8, 145-171.

Townsend, C.R. & Calow, P. (Eds) (1981) Physiological Ecology : an Evolutionary Approach to Resource Use. Blackwell Scientific Publications, Oxford.

Treton, J.A. & Coutois, Y. (1982) Correlation between DNA excision repair and mammalian lifespan in lens epithelial cells. Cell Biol. Int. Rep., 6, 253-260.

Williams, G.C. (1957) Pleiotropy, natural selection and the evolution of senescence. Evolution, 11, 398-411.

Woolhouse, H.W. (1967) The nature of senescence in plants. In: Aspects of the Biology of Ageing: Symposia of the Society of Experimental Biology, No. XXI, Ed. H.W. Woolhouse, pp. 179-213. Cambridge University Press, Cambridge.

Wynne-Edwards, V.C. (1962) Animal Dispersion in Relation to Social Behaviour. Oliver and Boyd, Edinburgh.

EVOLUTION OF THE BREADTH OF BIOCHEMICAL ADAPTATION

M. Lynch
W. Gabriel

INTRODUCTION

The existence of variation at the gene level is now accepted to be ubiquitous among species, although the extent of variation is known to differ among species (Nei, 1975; Ayala, 1976; Nei & Koehn, 1983; Nei & Graur, 1984). Much, but not all, of this variation can be explained by the neutral theory of molecular evolution (Kimura, 1983) which assumes that the dominant forces underlying the dynamics of gene frequency change are mutation and the random drift resulting from finite population size. While the neutral theory has been particularly successful in explaining the approximate constancy of the long-term rate of gene substitution, there are many short-term and localized properties of structural gene polymorphisms that can only be explained by selection, and a consensus is gradually emerging that the fit of the neutral theory to existing data is improved when the theory is modified to incorporate a second kind of drift due to random variation in selection intensity (Matsuda & Gojobori, 1979; Takahata & Kimura, 1979; Takahata, 1981; Nei & Graur, 1984).

These recent developments are of interest since a very substantial amount of mathematical theory on the maintenance of genetic polymorphisms via spatial and temporal heterogeneity in the environment has been formulated, largely as an alternative to the neutral theory (for comprehensive reviews and syntheses see Karlin & Lieberman, 1974; Christiansen & Feldman, 1975; Felsenstein, 1976; Hedrick et al., 1976; Gillespie, 1978). The object of most of this theory has been to demonstrate how environmental variation can serve as a form of balancing selection, thereby preserving genetic polymorphisms. With the exception of the studies of Takahata & Kimura (1979); Takahata (1981) and Tier (1981), most of this theory either does not incorporate mutation or does so only in a very restrictive manner (two alleles with reversible mutation), and therefore has very little to contribute to our understanding of the evolution of biochemical adaptation. That is, while the theory may explain the degree of heterozygosity maintained when alleles are exposed to different schedules and spatial patterns of selection intensity, it does not consider the underlying determinants of an allele's sensitivity to environmental fluctuations.

In addition to its potential role in promoting polymorphisms, environmental heterogeneity may also select for genes whose biochemical products can maintain their functional integrity in the face of such variation. However, perhaps because environmental heterogeneity comes in many forms (spatial and temporal variance, both within and between generations, and in the genetic as well as the environmental background) and is generally subjectively defined, little mathematical theory or empirical work has been focused on the problem of breadth of biochemical adaptation. Only a verbal argument presented by Ayala and Valentine (Ayala et al., 1975; Valentine, 1976; Ayala & Valentine, 1979) bears directly on the issue.

Drawing from the fitness set theory of Levins (1968) as well as from considerable electrophoretic data that indicate a positive correlation between the stability of trophic resources and the level of genetic variation in pelagic and benthic marine invertebrates, the Ayala-Valentine hypothesis states that functionally-broad alleles that encode for a highly generalised phenotype are strongly favoured in temporally variable environments. Narrowly-adapted alleles are purported to be strongly selected against under these conditions, thereby resulting in a high degree of homozygosity. In more temporally stable environments, the spatial component of heterogeneity is thought to take precedence such that microhabitat specialization results in the maintenance of a variety of narrowly-adapted alleles through balancing selection. The Ayala-Valentine hypothesis is unique in its focus on both the properties of loci (polymorphism and heterozygosity) and genes (functional breadth), and it has subsequently been invoked in modified form to explain patterns of genetic variation in marine decapods (Nelson & Hedgecock, 1980) and marine fishes (Smith & Fujio, 1982).

The hypothesis can also be criticised on a number of grounds (Soule, 1976; Nei & Graur, 1984). First, there are a number of species, such as the cave fish Astyanax mexicanus (Avise & Selander, 1972) and fossorial mammals (Nevo et al., 1974), which are thought to live in highly stable environments but have exceptionally low levels of genetic variation. Any criticism on these grounds is weak, since it can always be argued that a small effective population size is responsible for the reduced amount of heterozygosity. A second and more serious problem is the subjective nature of the hypothesis. Unless the level of environmental heterogeneity can be objectively defined from the standpoint of the organism, there is little hope for testing the hypothesis. Finally, it is not clear that the verbal reasoning of Ayala & Valentine has led to the correct prediction. For example, it is not clear why microhabitat specialization cannot be equally or more pronounced in temporally variable environments than in stable ones. The existing theoretical work on the relation of environmental heterogeneity to heterozygosity alone is exceedingly complex, and with slight changes in assumptions, different investigators have often reached radically different conclusions. Modification of the existing theory to allow for the evolution of environmental sensitivity at the gene level is likely to further complicate the theory for genetic variability to an even greater degree.

Despite the shortage of attention that the problem of breadth of biochemical adaptation has received from evolutionary biologists, there is no *a priori* reason to expect it to be any less important as a means of genetic adaptation to variable environments than the maintenance of heterozygosity. Indeed, several empirical studies suggest that something other than the promotion of polymorphisms must occur when populations are exposed to increasing levels of environmental heterogeneity. While Powell (1971), Powell & Wistrand (1978) and McDonald & Ayala (1974) all found strikingly higher levels of genetic variance in laboratory populations of *Drosophila* exposed to spatial and temporal heterogeneity compared to controls, more recent studies (Minawa & Birley (1978); Mackay (1980), (1981); Haley & Birley (1983); Zirkle & Riddle (1983) have obtained either mixed or contrary results. Moreover, Nei (1980) has pointed out that, even in the presence of environmental heterogeneity, heterozygosity in the populations of Powell (1971), Powell & Wistrand (1978), and McDonald & Ayala (1974) was eroded more rapidly than could be accounted by random sampling drift; i.e. even in the studies with results most concordant with theoretical predictions, environmental heterogeneity enhanced the rate of loss of genetic variance relative to expectations under selective neutrality. Clearly, the problem of what environmental heterogeneity selects for is far from resolved.

Although we recognize the importance of genetic variation in heterogeneous environments, we will not concern ourselves with this issue here, focusing instead on the development of a theory for the evolution of the breadth of adaptation at the gene level. We therefore address the Ayala-Valentine hypothesis only in part. Much of what follows will take on an adaptational tone since we are primarily concerned with identifying the optimal level of functional flexibility of an allele in an effort to formalize the argument of Ayala and Valentine. Elsewhere, we will examine the extent to which the intensity of selection for alleles, and hence the likelihood of the evolution of the optimum and of the maintenance of heterozygosity, is modified by the level of environmental variation (Lynch & Gabriel, in prep.). We will also present our mathematical derivations in detail elsewhere, restricting our attention here only to the most fundamental definitions and formulae.

THE GENIC FITNESS FUNCTION

In order to explore the consequences of environmental heterogeneity for the evolution of gene properties, we require a theory that explicitly links the fitness of an allele to its environment. In the following, we define the genic fitness function, $w(g_1,g_2|\phi)$, as the expected fitness of an allele over a continuous environmental gradient, ϕ . It is easiest to think of ϕ as a density-independent parameter such as temperature, and the allele as a gene encoding for an enzyme. Associated with any allele will be a mean environmental optimum (g_1) and a measure of functional flexibility (g_2). A precise definition of g_2 will follow below, but in effect it is a measure of the equitability of the fitness of an allele over the environmental gradient. It is important to note that g_1 and g_2 are not in vitro properties of an allele. On the contrary, they are functions of the integrated phenotype, being defined as the average environmental optimum and breadth of all individuals containing the allele.

In the derivation of a fitness estimate for allele (g_1,g_2) two sources of variation must be taken into consideration. First, because of mutation, recombination, segregation, and gene flow, any gene is likely to be found in a multitude of genetic backgrounds. Moreover, variation in the environment is likely to further magnify the diversity of phenotypes within which a gene resides. The environment within which an individual develops may have long-lasting effects on the phenotype (Falconer, 1981), and short-term changes may be accomplished by physiological acclimation or behavioral modification (Hochachka & Somero, 1973; Ricklefs, 1979). Thus, while (g_1,g_2) is defined to be the expected phenotype of an individual with the allele, the actual genic fitness function is determined by the conditional phenotype distribution $p(z_1,z_2 \mid g_1,g_2)$, as well as by the phenotypic fitness function, $w(z_1,z_2 \mid \phi)$ (Fig. 1), such that:

$$w(g_1,g_2|\phi) = \int_0^{+\infty} \int_{-\infty}^{+\infty} w(z_1,z_2|\phi).p(z_1,z_2|g_1,g_2).dz_1 dz_2 \qquad (1)$$

The second source of variation that influences the fitness of a gene is the variance in the environmental parameter ϕ due to spatial heterogeneity and temporal fluctuations. We will take this topic up in some detail in the following section. First, however, we consider the fundamental relationship of the fitness of a gene to the environment.

We assume that the environmental state, ϕ , is measured on a scale such that the phenotypic fitness function is normalised,

$$w(z_1,z_2|\phi) = (2\pi z_2)^{-1/2} \exp[-(z_1-\phi)^2/2z_2] \qquad (2)$$

For an individual with phenotype (z_1,z_2), z_1 is the environmental state in which fitness is maximized, and z_2, the "variance" of the fitness function. Note that z_1 and z_2 are measured on different scales, a small inconvenience that can be eliminated by square root transformation of the latter. Throughout this paper will refer to $\sqrt{z_2}$ as the environmental breadth of a <u>phenotype</u>. Note also that a cost to being a generalist is implicit in equation (2), since any increase in $\sqrt{z_2}$ results in a reduction of fitness in the optimal environment while increasing fitness in more extreme environments. Such a cost is implicit in the Ayala-Valentine hypothesis and throughout the evolutionary ecological literature (MacArthur, 1972; Pianka, 1978), although as emphasised by Futuyma <u>et al</u>., (1984) and Huey & Hertz (1984) empirical tests of the idea are almost totally lacking.

We further assume that the genes that contribute to an individual's environmental optimum and breadth have additive effects both within and between loci (for supportive arguments for enzymatic loci, see Gillespie & Langley, 1974 and Kacser & Burns, 1981) and that the conditional phenotype distributions for z_1 and z_2 are independent so that

$$p(z_1,z_2|g_1,g_2) = p(z_1|g_1).p(z_2|g_2) \qquad (3)$$

The conditional phenotype distribution for the environmental optimum, $p(z_1|g_1)$, is taken to be normal with mean g_1, and variance V'_{T1}. V'_{T1} includes the variance in environmental effects that contribute to the optimum, V_{E1}, and all of the genetic variance for the trait conditional on one copy of the gene being present at the locus. Thus,

Fig. 1. Mean fitness function for a gene (solid line) and for a few of the phenotypes within which it is found.

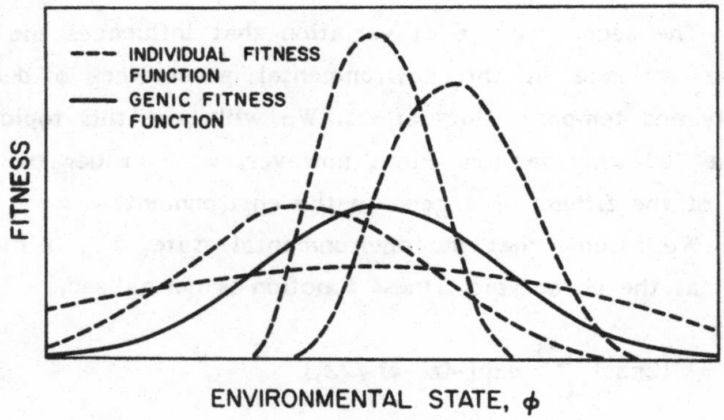

Fig. 2. Three conditional phenotype distributions for the square of environmental breadth (z_2), shapes of the distributions for V'_{T2} = 1, 10, 100, and shapes of the genic fitness function when applied to equation (1) with g_1=0, g_2=5, and V'_{T1}=1. For gamma-2 and inverse Gaussian distributed z_2, the genic fitness function was obtained by numerical integration (Gauss-Laguerre quadrature), whereas the approximate analytical solution (equation (5)) was used in the case of the beta distribution of the second kind.

ⓐ

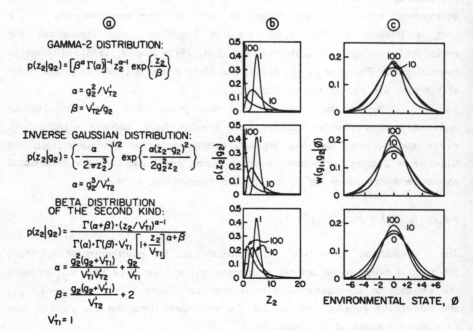

GAMMA-2 DISTRIBUTION:

$$p(z_2|g_2) = \left[\beta^{\alpha}\,\Gamma(\alpha)\right]^{-1} z_2^{\alpha-1} \exp\left\{-\frac{z_2}{\beta}\right\}$$

$$\alpha = g_2^2/V'_{T2}$$

$$\beta = V'_{T2}/g_2$$

INVERSE GAUSSIAN DISTRIBUTION:

$$p(z_2|g_2) = \left\{\frac{\alpha}{2\pi z_2^3}\right\}^{1/2} \exp\left\{-\frac{\alpha(z_2-g_2)^2}{2g_2^2 z_2}\right\}$$

$$\alpha = g_2^3/V'_{T2}$$

BETA DISTRIBUTION OF THE SECOND KIND:

$$p(z_2|g_2) = \frac{\Gamma(\alpha+\beta)\cdot(z_2/V'_{T1})^{\alpha-1}}{\Gamma(\alpha)\cdot\Gamma(\beta)\cdot V'_{T1}\left[1+\frac{z_2}{V'_{T1}}\right]^{\alpha+\beta}}$$

$$\alpha = \frac{g_2^2(g_2+V'_{T1})}{V'_{T1}V'_{T2}} + \frac{g_2}{V'_{T1}}$$

$$\beta = \frac{g_2(g_2+V'_{T1})}{V'_{T2}} + 2$$

$$V'_{T1} = 1$$

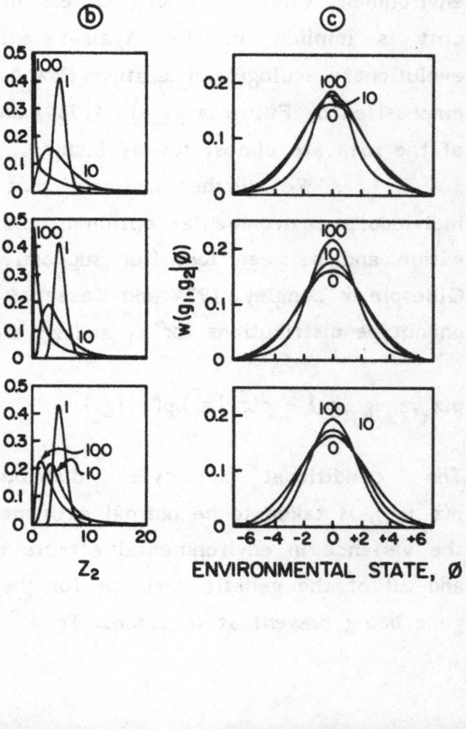

$$V'_{T1} = \rho_1 V_{G1} + V_{E1} \quad (4)$$

where V_{G1} is the total genetic variance for the optimum. Note that ρ_1 takes on a minimum value of 0.5 when no other loci encode for the environmental optimum, since only a single gene is free to vary. This low value may also be approached if the locus under consideration is in extreme linkage disequilibrium with other loci encoding for the environmental optimum. However, if linkage is not strong, $\rho_1 \to 1$ as the number of loci increases.

Since a variance cannot be <0, the conditional distribution for environmental breadth cannot be normal. Moreover, it does not seem biologicallly reasonable for z_2 to ever equal zero exactly. In order to separate genetic from environmental effects a distribution for which the variance is independent of the mean is desirable. Three distributions that meet these requirements are outlined in Fig. 2a. In each case the expectation of z_2 is g_2 and the variance of z_2 is V'_{T2} (defined in the same manner as V'_{T1}). All three distributions are very similar if the coefficient of variation $[(V'_{T2})^{1/2}/g_2] < 0.5$,and although they differ quantitatively for larger V'_{T2}, their qualitative behaviour is the same (Fig. 2b). When $(V'_{T2})^{1/2} \ll g_2$, $p(z_2|g_2)$ is approximately normal; but as V'_{T2} increases, the conditional distribution becomes increasingly asymmetrical with the mode approaching the origin. In the following analyses we rely on the beta distribution of the second kind for $p(z_2|g_2)$ as it is the only case for which we have been able to obtain an analytical solution for the genic fitness function.

The solution to equation (1) for normally distributed $p(z_1|g_1)$ and beta distributed $p(z_2|g_2)$ is somewhat involved and is presented elsewhere (Lynch & Gabriel, in press). It is sufficient to note here that if the scale is set such that $V'_{T1} = 1$, then provided $g_2 > 1$ (an assumption that is supported by analyses in the next section) and the coefficient of variation $[(V'_{T2})^{1/2}/g_2] < 1$ (a liberal upper limitfor most quantitative traits), the genic fitness function is closely approximated by

$$w(g_1,g_2|\phi) = (2\pi V')^{-1/2}\exp[-(g_1-\phi)^2/2V'] \quad (5)$$

where

$$V' = V'_{T1} + \frac{g_2\{g_2(g_2+V'_{T1}) + V'_{T2}\}}{g_2(g_2+V'_{T1}) + 2V'_{T2}} . \quad (6)$$

Thus, the genic fitness function is approximately normal with maximum fitness in environment $\phi = g_1$, and variance V'. In the following we will refer to $\sqrt{V'}$ as the realised environmental breadth for a gene. Note that for the special case in which $V'_{T2} = 0$, equation (5) reduces to the exact solution of equation (1) under those circumstances:

$$w(g_1,g_2|\phi) = [2\pi(g_2+V'_{T1})]^{-1/2} \exp[-(g_1-\phi)^2/2(g_2+V'_{T1})] \quad (7)$$

and that at the upper limit of the applicability of the normal approximation, $[(V'_{T2})^{\frac{1}{2}}/g_2] \simeq 1$. the genic fitness function is approximately

$$w(g_1,g_2|\phi) = \{2\pi[(2g_2/3)+V'_{T1}]\}^{-1/2} \exp\{-(g_1-\phi)^2/2[(2g_2/3)+V'_{T1}]\} \quad (8)$$

Thus, depending on the genetic and environmental background, an allele's realised environmental breadth is expected to fall within the range of $[(2g_2/3)+V'_{T1}]$ and $[g_2+V'_{T1}]$.

In summary, V'_{T1} and V'_{T2} have conflicting effects on the genic fitness function. Variance for the environmental optimum always results in a flattening of the fitness function, thereby endowing an allele with functional flexibility, whereas variance for environmental breadth has the opposite effect. The latter effect is not unique to the beta distribution of the second kind (Fig. 2c).

SPATIAL AND TEMPORAL HETEROGENEITY

Having defined the fundamental relationship between genic fitness and ϕ, we now proceed to evaluate the realised fitness of an allele with properties (g_1,g_2) in environments characterised by different degrees of spatial and temporal variation. For a population growing in discrete generations, we identify the mean environmental state over all microhabitats in generation t as ϕ_t. If the environment is spatially heterogeneous relative to the mobility of individuals, then it is likely that the actual mean environmental state experienced by an individual will be somewhat different from ϕ_t. We will take the mean environmental state experienced by individuals, ϕ_s, to be normally distributed with expectation ϕ_t and variance $V_{\phi s}$. Note that $V_{\phi s}$ is a measure of spatial heterogeneity perceived by the population. It is likely that $V_{\phi s}$ will be higher for a sedentary species than for a mobile species in the same environment. However, this need not always be the case, since dominance hierarchies in behaviourally sophisticated organisms might actually inflate the variance in mean environmental states experienced by different individuals.

The second source of environmental variation that we incorporate is the temporal variance in ϕ experienced by individuals within a generation, $V_{\phi tw}$. We will assume that the level of temporal heterogeneity experienced by individuals is independent of their mean environmental state, but make no assumptions regarding the temporal distribution of ϕ.

Before proceeding, it is worth considering the relationship of $V_{\phi s}$ and $V_{\phi tw}$ to the concept of environmental grain (Levins, 1968) which is frequently alluded to in the ecological genetic literature. An organism is said to perceive its environment as fine-grained if it passes through many patches in its life time. As individuals spend a greater proportion of their lives in a single microhabitat, the environment is said to be more coarse-grained. While the concept of grain is used in the context of spatial heterogeneity, it also has an element of temporal heterogeneity embedded in it. In our terminology, a fine-grained environment is one in which $V_{\phi s}$ is relatively low but $V_{\phi tw}$ is relatively high. That is, $\phi_s \simeq \phi_t$ for most individuals since they all pass through most patch types in their life times, but the movement between patches with different environmental states increases $V_{\phi tw}$. In the most fine-grained of environments, $V_{\phi s}$ will be essentially zero, but even in the most coarse-grained of environments, while $V_{\phi tw}$ will be reduced, it cannot be less than the temporal variance of ϕ which occurs within a microhabitat. Thus, the definition of environmental grain, which has heretofore been used primarily from a heuristic standpoint, is formalized by the use of $V_{\phi s}$ and $V_{\phi tw}$.

A more general interpretation of $V_{\phi s}$ and $V_{\phi tw}$ is to consider them to be the total variance in additive and multiplicative effects on fitness resulting from environmental heterogeneity. While we have defined the spatial component of environmental variation such that it has an additive influence on genic fitness, temporal variance in ϕ within generations influences fitness geometrically (as when daily probabilities of survival interact multiplicatively to determine annual survivorship). Using techniques that we present elsewhere (Lynch & Gabriel, in press), the fitness of allele (g_1,g_2) in generation t is found to be

$$w(g_1,g_2,t) = \{2\pi(V'+V_{\phi s})\}^{-1/2}.\exp\left\{-\frac{1}{2}\left[\frac{V_{\phi tw}}{V'} + \frac{(g_1-\phi_t)^2}{V'+V_{\phi s}}\right]\right\} \qquad (9)$$

Finally, in order to determine the relative long-term advantages of different alleles, we incorporate the variance in ϕ_t' between generations. In extending our analysis across generations, we adopt the geometric mean fitness of a gene as a measure of its relative success (Dempster, 1955). In order for us to make any analytical progress, it is necessary to assume constancy of the genetic background (ρ_1, ρ_2, V_{G1}, and V_{G2}) throughout the period of selection. Approximate constancy of these parameters can be expected to result from selection-mutation balance in an environment that

Fig. 3. The optimum g_2 of a gene product as a function of the spatial ($V_{\phi s}$), within-generation temporal ($V_{\phi tw}$), and between-generation temporal ($V_{\phi tb}$) variance. Solutions are for the case in which $g_1 = \bar{\phi}_t$ (i.e. the optimal environmental optimum has been attained), $V_{T1}' = 1$, and $V_{T2}' = 0$.

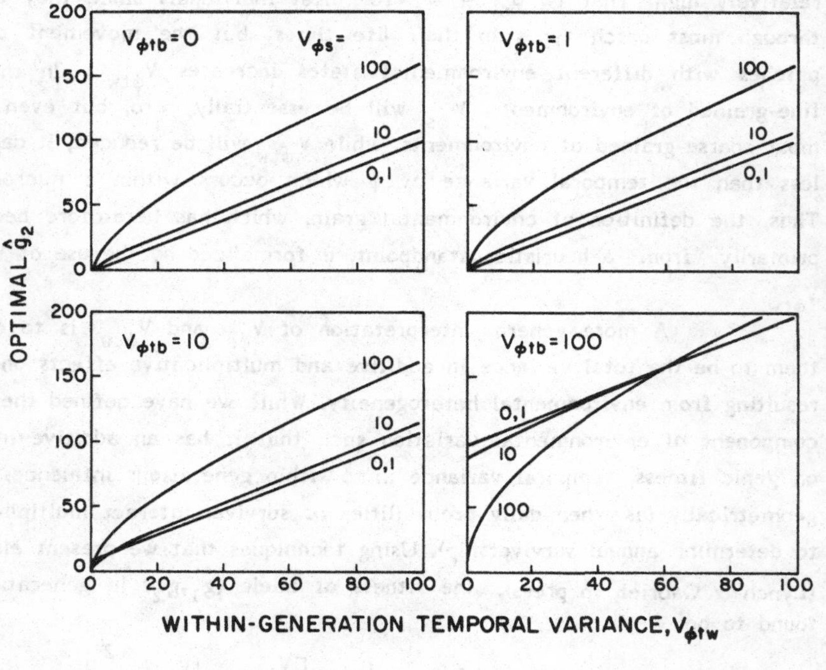

does not change between generations, but stochastic variation in ϕ_t must promote variance in the genetic background. Provided the distribution of ϕ_t is stochastically stable, we anticipate that stochastically stable distributions will also arise for ρ_1, ρ_2, V_{G1}, and V_{G2}, thereby preventing (or retarding) complete fixation or loss of alleles with different properties in effectively infinite populations (Gillespie, 1978). We hope to address this complicated issue in the future. For now we simply take the position that the expected frequency of an allele will be positively correlated with its geometric mean fitness as defined below.

Setting the environmental scale so that the long-term mean environmental state (average ϕ_t) is zero, and letting the variance in ϕ_t between generations be $V_{\phi tb}$, the geometric mean fitness of an allele with properties (g_1,g_2) is found using the techniques in Lynch & Gabriel (in press),

$$w(g_1,g_2) = \lim_{T\to\infty} \left[\prod_{t=1}^{T} w(g_1,g_2,t)^{1/T} \right]$$

$$= \{2\pi(V'+V_{\phi s})\}^{-1/2} \exp\left\{ -\frac{1}{2}\left[\frac{V_{\phi tw}}{V'} + \frac{g_1^2 + V_{\phi tb}}{V' + V_{\phi s}}\right]\right\} \qquad (10)$$

As in the case of within-generation temporal variance, this derivation makes no assumptions about the temporal distribution of ϕ_t between generations.

We are now in a position to evaluate the influence of the various forms of environmental variance on the optimal properties of an allele $(\hat{g}_1,\hat{g}_2)$. It is immediately clear that the environmental optimum that maximizes geometric mean fitness is $\hat{g}_1 = 0$, i.e., the long-term mean environmental state. Estimation of the optimal environmental breadth, $\hat{g}_2$, is less straight-forward because of the complexity of V'. However, the limits to $\hat{g}_2$, can be obtained quite readily by computer by noting from equations (7) and (8) that $[(2\hat{g}_2/3) + V'_{T1}] \leq V' \leq [\hat{g}_2 + V'_{T1}]$ and that from the standpoint of the individual, it is the realised environmental breadth (V') that must be optimized. The minimum value for $\hat{g}_2$, which arises when V'_{T2} is zero, is given in Fig. 3 as a function of $V_{\phi s}$, $V_{\phi tw}$, and $V_{\phi tb}$. The maximum $\hat{g}_2$, which we assume to be approached as $(V'_{T2})^{1/2}/g_2 \to 1$, is approximately 1.5 times the plotted values.

Several conclusions can be drawn from this analysis. First, in accordance with Ayala & Valentine's expectation, temporal variance in the environment always results in selection for more broadly adapted or "functionally flexible" alleles. However, the scale of temporal variance matters a great deal. In an extremely fine-grained environment ($V_{\phi s} = 0$),

$V_{\phi tw}$ and $V_{\phi tb}$ have identical influences on $\hat{g}_2$, but as $V_{\phi s}$ becomes large, the influence of the between-generation variance becomes negligible. The significance of temporal variation between generations is diminished in highly spatially heterogeneous environments because a relatively even distribution of environmental states is already present in different microhabitats, and a shift in $\overline{\phi}_t$ between generations does little to change it.

Second, and in partial agreement with Ayala & Valentine, spatial heterogeneity is not a sufficient condition for the evolution of broadly adapted alleles. In temporally invariant environments, functionally narrow alleles (those with the lowest possible g_2) will have the highest expected fitness. However, provided that the between-generation variance is not too large, spatial heterogeneity in temporally unstable environments accentuates selection for generalism in environments with higher $V_{\phi tw}$. This result is in conflict with the implicit assumption of the Ayala-Valentine hypothesis that spatial heterogeneity plays a diminishing role in molding adaptive genetic properties in temporally variable environments.

Finally, we note that if the between-generation component of temporal variance is much greater than the within-generation component, an inverse relationship may actually arise between $V_{\phi s}$ and $\hat{g}_2$. Such an effect is not predicted by the verbal hypothesis of Ayala & Valentine. It appears to result because a narrowly adapted allele in a spatially homogeneous environment is highly sensitive to environmental changes between generations, whereas the mean fitness of a more broadly adapted allele is relatively constant from generation to generation. In a highly spatially heterogeneous environment a specialist allele would nearly always be located in some favorable microhabitats even in generations with extreme ϕ_t. For example, if ϕ were temperature, then in a year with an extremely high temperature, a microhabitat that is on average excessively cool would have a temperature close the long-term average.

DISCUSSION

The sensitivity of a biochemical pathway to environmental fluctuations most likely evolves in response to two conflicting forces : the necessity of a generalised strategy in a heterogeneous environment, and the cost of evolving generalism (the "jack-of-all-trades is a master-of-none" argument). At the level of biochemical adaptation, environmental heterogeneity includes not only spatial and temporal variance external to the individual (our $V_{\phi s}$, $V_{\phi tw}$, and $V_{\phi tb}$), but also variation in the genetic environment resulting from a gene's residence in a number of genetic backgrounds (our $\rho_1 V_{G1}$ and $\rho_2 V_{G2}$) and variation in the developmental background induced by the environment (our V_{E1} and V_{E2}). We may, therefore, expect the evolution of biochemical/physiological strategies of organisms to be as much constrained by population genetic phenomena (degree of inbreeding, migration, linkage and chromosomal structure) as by extrinsic environmental factors.

The study of the breadth of biochemical/physiological adaptation from an evolutionary perspective is clearly in an early embryonic state. Indeed, we know of no explicit attempts to test the Ayala-Valentine hypothesis nor of any data sets to which our own theory may be applied. A substantial amount of comparative work has been done on the response of isozymic reactions to changes in the physical/chemical environment, but virtually all of this work has either been done _in vitro_ or in fixed genetic backgrounds (Hochachka & Somero, 1973; Koehn _et al._, 1983; Watt, 1985). Any attempt to measure g_1 and reconcile it with an adaptationist argument must realise that the relevant measures of biochemical properties are those obtained _in vivo_ rather than _in vitro_, a point recently emphasised by several biochemical geneticists (Middleton & Kacser, 1983; Powell & Amato, 1984; Watt, 1985), and that meaningful measures can only be obtained by examining an allele's properties in a full complement of genetic backgrounds rather than in a single artificially constructed background.

An important reason why an _in vivo_ measure may provide an inaccurate description of the _in vitro_ properties of an allele underlying a polygenic trait was pointed out by Lande (1976). The only constraint on a polygenic trait under stabilizing selection is that the aggregate effect of all constituent loci results in a phenotype near the optimum. Subject to this single constraint, the mean effects of constituent loci are free to change in an infinite number of ways via the interaction of drift, mutation, and selection.

In this first attempt at a formal theory for the breadth of biochemical adaptation we have made a number of assumptions regarding shapes of distributions, additivity of allelic effects, an effectively infinite number of possible allelic types, a cost to specialization, and zero covariance between the mean and variance of environmental states within microhabitats. Such assumptions, some of which have been rationalised above, have been necessary in order for us to make progress in the development of a theory that is mechanistic and couched in terms that are measurable in natural populations. Regardless of these assumptions, however, it is clear that at least seven types of variance in the genetic and environmental background have an influence on the optimal environmental breadth of an allele. These are summarised with their implications for the evolution of the breadth of biochemical adaptation in Table 1.

Perhaps the most striking result of our analysis is the conclusion that not all types of variation in the background of an allele encourage the evolution of broad adaptation ("functional flexibility" or "generalism") for the gene product. If the variance for the environmental optimum (V'_{T1}) is high, either because of high V_{E1} or $\rho_1 V_{G1}$, then some individuals containing the allele are likely to be at their optimal environmental state in most generations, and the cost of evolving generalism can be avoided. Moreover, as discussed above, depending on the temporal stability of the environment, spatial variance, caused by structural complexity of the environment and/or

Table 1. Components of variance in the background of an allele and the influence that they have on the evolution of breadth of biochemical adaptation.

Factor		Increased breadth of biochemical adaptation, $\sqrt{g_2}$, is favoured if the factor
$\rho_1 V_{G1}$,	conditional genetic variance for environmental optimum	decreases
V_{E1},	environmental variance for the environmental optimum	decreases
$\rho_2 V_{G2}$,	conditional genetic variance for environmental breadth	increases
V_{E2},	environmental variance for environmental breadth	increases
$V_{\phi s}$,	spatial component of the variance in environmental state as perceived by individuals	depends on $V_{\phi tw}$ and $V_{\phi tb}$
$V_{\phi tw}$,	within-generation component of temporal variance in environmental state as perceived by individuals	increases
$V_{\phi tb}$,	between-generation component of temporal variance in environmental state as perceived by individuals	increases

immobility of individuals, can sometimes select for reduced environmental breadth.

Although some of our conclusions may be quantitatively sensitive to changes in the assumptions underlying our model, our results clearly indicate that the complexity of the relationship between environmental heterogeneity and the evolution of genetic adaptations should not be taken lightly. Although Ayala and Valentine were not explicit in formulating their hypothesis that temporal instability of the environment results in selection for broadly adapted alleles, they appear to adopt $V_{\phi tw}$ as their measure of temporal heterogeneity. Our results indicate that the relation between $V_{\phi tw}$ and $\hat{g}_2$ in different species and/or populations will be highly dependent on the degree to which spatial heterogeneity and between-generation variation in ϕ_t are correlated with $V_{\phi tw}$.

The most appropriate test of our theory for a specific locus would involve the measurement of life-time fitness of many individuals known to contain the allele of interest, but otherwise randomly taken from a natural population. By performing such measures at various points along the environmental gradient and subsequently curve-fitting, estimates of g_1 and $\sqrt{V'}$ can be obtained for the allele. Thus, in the context of evolutionary theory, the measurement of the breadth of biochemical adaptation need not involve any biochemical analysis other than the electrophoresis needed to identify the genotypes of individuals. It does, however, require the analysis of many more individuals than an in vitro biochemical investigation.

Although the further partitioning of the realised environmental breadth, V', into its components (g_2, $\rho_1 V_{G1}$, $\rho_2 V_{G2}$, V_{E1}, and V_{E2}) would be of interest, it would also require extensive quantitative genetic analysis and for many purposes may be unnecessary. Simple estimates of $\sqrt{V'}$ are relevant to our theory. For while we have focused on the optimization of g_2 in this paper, the evolution of g_2 is actually determined by the more fundamental constraint that $\sqrt{V'}$ be optimized. Since we have scaled our parameters in this paper such that $V'_{T1} = 1$, the optimal values for V' are obtainable from Fig. 3 (for the case $V'_{T2} = 0$) by simply adding 1 to $\hat{g}_2$.

We are very grateful to S. Portnoy for the suggestion of the use of the beta distribution of the second kind and to J. Gresey, A. Lee, and S. Schiller for analytical assistance. Supported by National Science Foundation grants BSR 83-06072 and SUB U MICHX98764 and a fellowship from the Max Planck Society to ML.

REFERENCES

Avise, J.C. & Selander, R.K. (1972) Evolutionary genetics of cave-dwelling fishes of the genus Astyanax. Evolution, 26, 1-19.

Ayala, F.J. (Ed.) (1976) Molecular Evolution. Sinauer Assocs. Sunderland, Mass.

Ayala, F.J. et al. (1975) An electrophoretic study of the antarctic zooplankter Euphausia superba. Limnol. Oceanogr., 20, 635-639.

Ayala, F.J. & Valentine, J.W. (1979) Genetic variability in the pelagic environment : a paradox? Ecology, 60, 24-29.

Christiansen, F.B. & Feldman, M.W. (1975) Subdivided populations : a review of the one- and two-locus deterministic theory. Theor. Pop. Biol., 7, 13-38.

Dempster, E.R. (1955) Maintenance of genetic heterogeneity. Cold Spring Harbor Symp. Quant. Biol., 20, 25-32.

Falconer, D.S. (1981) Introduction to Quantitative Genetics. Longman Inc. New York.

Felsenstein, J. (1976) The theoretical population genetics of variable selection and migration. Ann.Rev.Genet., 10, 253-280.

Futuyma, D.J. et al. (1984) Adaptation to host plants in the fall cankerworm (Alsophila pometaria) and its bearing on the evolution of host affiliation in phytophagous insects. Am.Nat., 123, 287-296.

Gillespie, J.H. (1978) A general model to account for enzyme variation in natural populations. V. The SAS-CFF model. Theor.Pop.Biol., 14, 1-45.

Gillespie, J.H. & Langley, C.H. (1974) A general model to account for enzyme variation in natural populations. Genetics, 76, 837-848.

Haley, C.S. & Birley, A.J. (1983) The genetical response to natural selection by varied environments. II. Observations on replicate populations in spatially varied laboratory environments. Heredity, 51, 581-606.

Hedrick, P.W. et al. (1976) Genetic polymorphism in heterogeneous environments. Ann.Rev.Ecol.Syst., 7, 1-32.

Hochachka, P.W. & Somero, G.N. (1973) Strategies of Biochemical Adaptation. W.B. Saunders Co., Philadelphia, PA.

Huey, R.B. & Hertz, P.E. (1984) Is a jack-of-all-temperatures a master of none? Evolution, 38, 441-443.

Karlin, S. & Lieberman, U. (1974) Random temporal variation in selection intensities: case of large population size. Theor.Pop.Biol., 6, 355-382.

Kacser, H. & Burns, J.A. (1981) The molecular basis of dominance. Genetics, 97, 639-666.

Kimura, M. (1983) The Neutral Theory of Molecular Evolution. Cambridge Univ. Press, New York.

Koehn, R.K. et al. (1983) Enzyme polymorphism and natural selection. In: Evolution of genes and proteins, Eds. M. Nei & R.K. Koehn, pp. 115-136. Sinauer Assocs., Sunderland, Mass.

Lande, R. (1976) The maintenance of genetic variability by mutation in a polygenic character with linked loci. Genet.Res., 26, 221-235.

Levins, R. (1968) Evolution in Changing Environments. Princeton Univ. Press, Princeton, NJ.

Lynch, M. & Gabriel, W. (1987) Environmental tolerance. Am. Nat. (in press).

MacArthur, R.H. (1972) Geographical Ecology. Harper and Row, Publ., New York.

Mackay, T.F.C. (1980) Genetic variance, fitness, and homeostasis in varying environments: an experimental check of the theory. Evolution, 34, 1219-1222.

Mackay, T.F.C. (1981) Genetic variation in varying environments. Genet. Res., 37, 79-93.

Matsuda, H. & Gojobori, T. (1979) Protein polymorphism and fluctuation of environments. Ad.Biophys., 12, 53-99.
McDonald, J.F. & Ayala, F.J. (1974) Genetic response to environmental heterogeneity. Nature, 250, 572-574.
Middleton, R.J. & Kacser, H. (1983) Enzyme variation, metabolic flux and fitness: alcohol dehydrogenase in Drosophila melanogaster. Genetics, 105, 633-650.
Minawa, A. & Birley, A.J. (1978) The genetical response to natural selection by varied environments. Heredity, 40, 39-50.
Nei, M. (1975) Molecular Population Genetics and Evolution. Amsterdam, N. Holland.
Nei, M. (1980) Stochastic theory of population genetics and evolution. In: Vito Volterra symposium on mathematical models in biology. (Lecture notes in biomathematics 39), Ed. C. Barigozzi, pp. 17-47. Springer-Verlag., Berlin.
Nei, M. & Graur, D. (1984) Extent of protein polymorphism and the neutral mutation theory. Evol.Biol., 17, 73-118.
Nei, M. & Koehn, R.K. (Eds.) (1983) Evolution of Genes and Proteins. Sinauer Assocs., Sunderland, Mass.
Nelson, K. & Hedgecock, D. (1980) Enzyme polymorphism and adaptive strategy in the decapod Crustacea. Am.Nat., 116, 238-280.
Nevo, E. et al. (1974) Genetic variation, selection and speciation in Thomomys talpoides pocket gophers. Evolution, 28, 1-23.
Pianka, E.R. (1978) Evolutionary Ecology. 2nd Edn. Harper and Row, New York.
Powell, J.R. (1971) Genetic polymorphisms in varied environments. Science, 174, 1035-1036.
Powell, J.R. & Amato, G.D. (1984) Population genetics of Drosophila amylase. V. Genetic background and selection on different carbohydrates. Genetics, 106, 625-629.
Powell, J.R. & Wistrand, H. (1978) The effect of heterogeneous environments and a competitor on genetic variation in Drosophila. Am.Nat., 112, 935-947.
Ricklefs, R. (1979) Ecology. Chiron Press, New York.
Smith, P.J. & Fujio, Y. (1982) Genetic variation in marine teleosts: high variability in habitat specialists and low variability in habitat generalists. Mar.Biol., 69, 7-20.
Soule, M. (1976) Allozyme variation: its determination in space and time. In: Molecular Evolution, Ed. F.J. Ayala, pp 60-77 Sinauer Assocs., Sunderland, Mass.
Takahata, N. (1981) Genetic variability and rate of gene substitution in a finite population under mutation and fluctuating selection. Genetics, 98, 427-440.
Takahata, N. & Kimura, M. (1979) Genetic variability maintained in a finite population under mutation and autocorrelated random fluctuation of selection intensity. Proc. Natl. Acad. Sci. U.S.A., 76, 5813-5817.
Tier, C. (1981) An analysis of neutral-alleles and variable-environment diffusion models. J.Math.Biol., 12, 53-71.
Valentine, J.W. (1976) Genetic strategies of adaptation. In: Molecular Evolution, Ed. F.J. Ayala, pp. 78-94. Sinauer Assocs., Sunderland, Mass.
Watt, W.B. (1985) Bioenergetics and evolutionary genetics: opportunities for new synthesis. Am.Nat., 125, 118-143.
Zirkle, D.F. & Riddle, R.A. (1983) Quantitative genetic response to environmental heterogeneity in Tribolium confusum. Evolution, 37, 637-638.

EVOLUTION FROM THE VIEWPOINT OF ESCHERICHIA COLI

A.L. Koch

INTRODUCTION

Ferreting the evolutionary path giving rise to any organism is both sleuthing at its best and science fiction at its worst. But this is now possible for organisms that do not leave a fossil record. Here we will focus on Escherichia coli because the trail can now be followed in the sequence of nucleic acids, in metabolic pathways, and in our understanding of astronomical, geological, and biological history of that part of the universe that is not E. coli.

THE FIRST CELLS

A key point, possibly the most critical point, in the evolution of life was the development of a form that was subject to Darwinian selection. This first life form had to be capable of replication, of mutation (and inheriting those mutations), and of doing something so that selection, controlled by the external environment, could favour the better-adapted form. Once such a mechanism had occurred, adaptive evolution could make forms ever more complex, more sophisticated, and adapted to an ever-widening range of habitats. So I would like first to focus on the problems that must have been present at the time of origin of the first creature.

The cornerstone of evolutionary theory is that selection has to directly favour or to directly discriminate against the genetic element itself. This requires that life, even at the start, had to be cellular (Koch, 1984, 1985). Only if a life form is cellular and capable of dividing would a favourable mutation be in a position to favour itself and not to favour all less adapted genes of the same kind. If there were no cell boundaries, the less favourable forms would be favoured as well and selection would be slowed. The rule that the propagule is selected on its own merits had to remain inviolable until sufficiently sophisticated life forms arose and sociobiological considerations became important. This means that cell membranes, cell membrane growth, and cell division had to be very early aspects of the biology on any planet where chemically-based life was to evolve. (If life based on solid state physics were possible, similar constraints would be necessary.) But how soon in evolution can we imagine the development of cell growth and division. This process is tremendously complicated in all modern organisms.

In particular, could it have developed BP: before protein? Only when protein synthesis had been developed and refined could sufficiently complex structures be formed to do the precise jobs executed by the enzymes and regulatory proteins of all extant organisms. Is there a simple, crude way that would allow cells to grow, divide, and to achieve cell separation? I think there is (Koch, 1985). No doubt, in the primordial ooze, there were phospholipid-like molecules. It has been suggested by many workers that liposomes could have existed in the primitive oceans made by wave action. But such mechanically constructed vesicles do not have the property that allows them to be capable of indefinite division and enlargement in a way that would serve to separate and disperse genetic elements. But if the first living organism had, as one of its few genetic characteristics, an ability to favour the completion or formation of a phospholipid-like molecule, then I think that cell division could occur in a very natural, spontaneous way.

Fig. 1 shows the hypothetical life cycle of an early cell consisting of a phospholipid-like vesicle that forms phospholipid molecules internally. (This could have been due to a catalytic function based possibly on the secondary and tertiary structure of a nucleic acid). These de novo phospholipid molecules would be more hydrophobic than their precursors and would enter the inner leaflet of the phospholipid bilayer and consequently would favour invaginations of that leaflet; e.g. see upper right Fig. 1. As such invaginations continued to take place, a cell would be partitioned into smaller cells, still fused together. Consequently, addition to the inner leaflet has solved one of the requirements needed for the division process. A second function is needed, but this functional requirement is quite non-specific: it could be any way that coupled energy available in its environment into increasing the total number of molecules inside the vesicle. That would be enough to force the cell volume to grow and the pressure inside the cell to increase. As the osmotic pressure increased inside the vesicle, the hydrostatic pressure would create a tension that would tend to favour a transition that is ordinarily very rare; namely, the flipping of a phospholipid molecule from the inner leaflet to the outer leaflet. This would allow, as shown in the lower

Fig. 1. Cell division due to internal biosynthesis of phospholipids. A phospholipid vesicle that internally completes phospholipids is depicted. Proceeding from the upper left, as the phospholipids are formed they enter the inner leaflet of the bilayer and thus form partitions. Accumulation of substance raises the hydrostatic pressure within the vesicle and leads to an increase in flipping of the molecules to allow the outer leaflet to enlarge. This spontaneously leads to division and separation. Reproduced from Koch (1985) by permission.

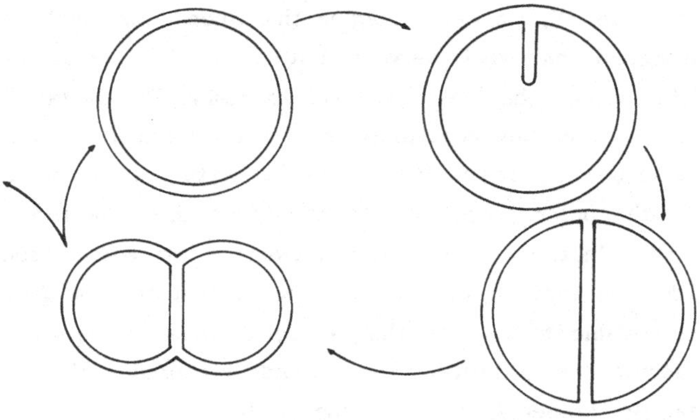

Fig. 2. A possible mechanism for energy generation in the most primitive cell. The hypothetical mechanisms facilitates a redox reaction, creates a proton gradient, pumps phosphate and generates high-energy phosphate bonds. Reproduced from Koch (1985) by permission.

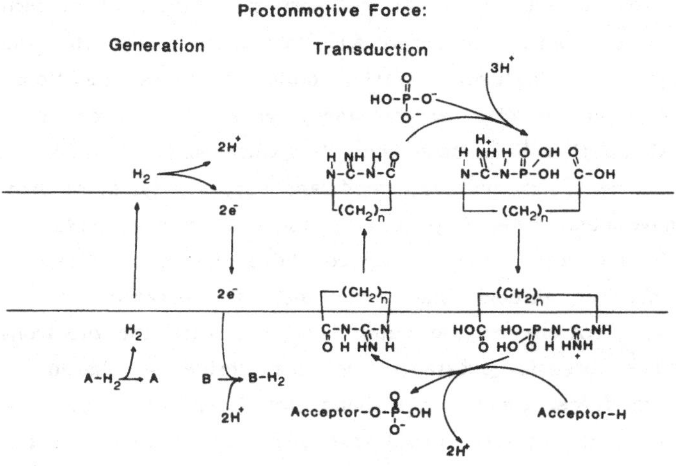

left, the splitting of the outer membrane and the continued increase in the volume contained within the phospholipid bounding membranes. The end product of phospholipid completion and increase in colligative particles would be the development of many separate cells that then could be dispersed. Such cell division may be seen in certain modern micro-organisms found in particular habitats; it can be seen in certain mutants constructed by the molecular biologist.

Now turning attention to the energy-coupling machinery of the first cell. For life BP, we must imagine that there were molecules available in the environment that would serve as natural couplers for bioenergetics. The source of the energy then, as now, had to reside in spontaneous processes driven by non-equilibrium conditions in the environment. Some modern-day organisms develop a protonmotive force by virtue of separating hydrogen molecules in the outer aspect of the membrane into electrons and protons, conducting the electrons through the membrane to reduce some chemical species inside, and thus, by deficit, creating a protonmotive force. Similarly, it would be possible to imagine that a protonmotive force could have been developed across the primitive phospholipid membrane, if the cell had a second catalytic property; namely, the ability to reduce some generally available molecule at the inside surface of the membrane. Possibly this asymmetry of reduction was because of local concentrations or because of the asymmetry of location of catalytic activity.

But simple transfer and reduction of electrons does not of itself force any energy into the biosynthetic processes of the cell; instead, something that is akin to active transport mechanisms of modern organisms and/or akin to ATPase that forms ATP in modern prokaryotes and organelles of eukaryotes is needed. Clearly, both of these functions are very sophisticated and revolve around many very complicated and specialised proteins. Could that have been done BP? Once again, I think that it could have been done - very poorly, indeed, and very slowly. More strongly, I feel it must have been done. A possibility, taken from Koch (1985), is shown in Fig. 2. On the left I have imagined the situation described above where hydrogen diffuses through the membrane, is separated to its component protons and electrons on the outer face, and then the electrons return to reduce some organic substance on the inside - thereby creating a protonmotive force. On the right-hand side I have pictured a possible way that the protonmotive force could have been coupled to favour the growth of the cell. While this is even more speculative, it does outline the necessary structural constraints. What is shown is a molecule akin to creatinine that on the outer face of the membrane reacts with inorganic phosphate and binds protons in such a way that the anhydride bond that forms the ring in the creatinine is transformed into an anhydride bond that is analogous to that in creatine phosphate.

The necessary requirement for coupling is that such a carrier molecule be capable of folding itself around the hydrophilic and electronegative substance to be translocated in the way that potassium is surrounded by the antibiotic valinomycin, then the complex with its hydrophobic exterior can pass through the membrane. Granting that this could be done either as indicated in Fig. 2 or in some equivalent way with naturally occurring materials present in the abiotic environment, then phosphate would bind on the outside, be carried to the inside and, in fact, be carried in a form that biochemists call a "high energy phosphate bond". In turn, on the innerface, the phosphate could be transferred to some acceptor, being in that act accumulated within the cell. In this transfer the anhydride form of the creatinine analogue is reformed and this, in turn, can transfer across the membrane to complete the cycle. This cycle is one that makes high-energy phosphate bonds and pumps phosphate and incidental to this primitive biochemistry, would do the job of forcing the volume growth that leads to cell division and separation. The latter, in turn, allowed the biosynthetic apparatus to develop and become precise; the organism to grow, and to genetically adapt and evolve.

ENERGY-COUPLING IMPROVEMENTS INCREASED THE WORLD CARRYING CAPACITY

Shortly after the beginning of life on this planet, growth must have been slow and the total biomass low. Even though evolution was consequently slow, mutation and growth no doubt produced organisms to exploit and deplete the environment. It can be assumed that a point was reached such that the planet was close to its carrying capacity, limited by dependence on the then current rate of abiotic generation of exploitable biochemicals. Further increase in biomass required the development of ways to extract energy from the environment into the metabolic systems. Based on the recent studies of Woese et al., (1985) on the sequences of 16S ribosomal RNA, there appear to be three deep divisions splitting the phylogeny of the prokaryotes. I suggest that these correspond to three radically different energy extraction procedures: (i) carbon dioxide reduction with available reductants, as carried out by modern representatives of the methanogenic archaebacteria; (ii) reduction of geologically released sulphate with available reductants, as in the modern Desulfovibrio and Desulfotomaculum; and (iii) cyclic phosphorylation, as in the modern, purple, non-sulphur photosynthetic bacteria. The last two categories correspond to the Gram-positive and Gram-negative bacteria, respectively.

From the point of view of a general microbiologist, both the phylogeny of the Woese school and my suggestion for the fundamental reason for the evolutionary splits are inconsistent with much detailed knowledge of modern prokaryotes. It is now impossible to evaluate the effects of more recent lateral transfers of genes, covergent evolution and loss of genetic information. But clearly these factors have blurred the original patterns of evolution.

Each of these advances would have permitted a great increase in world biomass and led to three radiations that gave rise to many of the further subdivisions we see today in the prokaryotes. But the geological limitations in sites where both the oxidant and reductant would have been simultaneously present would still have greatly limited the world biomass below today's value. With the development of oxygenic photosynthesis, the biomass would have increased greatly because the restrictions in habitat permitting growth would have been lessened. While the later development of terrestrial plants would have also produced an important increase in biomass, oxygenic photosynthesis was the development that permitted a great increase in biomass. This, in turn, first allowed certain organisms to become obligate utilizers of biomass. And that in turn, created niches and niches-upon-niches.

E. coli, in the view of Woese et al., (1985), descended from photosynthetic autotrophs, and in this way should have been endowed with metabolic pathways for the synthesis of amino acids, nucleic acids, bases, etc., as might be expected of an emeritus autotroph. It belongs to the branch of the purple bacteria designated by Woese and colleagues as gamma-3. This branch includes the Xanthomonas group, fluorescent pseudomonads, Acinetobacter, Oceanospirilla, Bacterioides mylophilus, Alteromonas putrefaciens, Vibrio, Aeromonas hydrophilia, and Pasteurella multocida. The latter two are the nearest neighbours of the enterics in the Woese classification. At closer range, E. coli has much in common with Klebsiella aerogenes and, of course, with Salmonella.

Why have most of these organisms lost the photosynthetic capabilities of their purple bacteria forbears? Is cyclic photophosphorylation no longer the boon it once was? Its disadvantages, in addition to its basic requirement for light and a reductant for growth, arose in some part because there were larger, more rapidly growing, and more elevated organisms shading the physical location of these bacteria. The remaining purple bacteria probably then altered their pigments to use the infrared radiant energy that still would reach them. I surmise that although many of the original purple bacteria had become autotrophs, at the same time they honed their catabolic pathways so that heterotrophic metabolism became an effective alternate way to cope 24 hours a day instead of just at night. After oxygenic photosynthesis became highly productive there was a continuing renewable source of energy and carbon at about the average oxidation states of living matter. By the catabolism of biologically produced matter, either with or without oxygen and either as an energy source or building blocks, the deletion of photosynthesis became practical and closer to the rule when the entire group is considered.

E. coli ECOLOGY

E. coli largely exploits a limited set of resources available in the gastrointestinal tract of mammals. It had to wait almost until the middle Jurassic to become tuned to that then-new niche. Possibly the ancestor of E. coli had adapted to the intestines of amphibians and reptiles; E. coli can be isolated from these habitats today. It also has become partially specialised in that E. coli is more common in mammalian intestines than in the intestines of birds (unless they are birds of prey).

The niche of E. coli is that of a commensal organism in the mammalian gut (Hartl & Dykhuizen, 1984; Koch, 1971, 1976, 1985; Koch & Schaechter, 1985). Most of the time, it neither helps nor hinders its host. Occasionally it can cause disease, and it aids its host by the production of various vitamins, such as vitamin K, but these are incidental. The niche is only for an organism that can secure certain low molecular weight nutrients before other more numerous organisms or the host itself can get at them. As such, it is a fringe organism that is seldom highly successful. One of its first attributes is its small size and its ability to take up and consume certain chemical molecules present at very low concentrations. It must have an ability to convert any of the organic compounds that it can consume into all of its constituents. Wild-type strains are capable of growing in the laboratory in very simple media, and it can be presumed that this ability is important in the wild state when it exists for long periods in a state of chronic starvation; it must survive and grow on sparse, qualitatively and quantitatively-limited resources that come into its vicinity.

Occasionally nutrients will be present in adequate amounts and any candidate organism must be able to quickly switch from a chronic starvation mode into the rapid growth mode. Its long-term survival depends on being able to seize such opportunities very quickly when they come along, consume the new-found resource and multiply very rapidly indeed (Koch, 1971). Wild-type strains, as obtained from nature, can typically grow with doubling times as fast as 16 min in a very rich medium at 37°C (Koch, 1980). (Note that most of the strains now present in laboratories have been subject to enough cycles of mutagenesis so that 20 min is the usual fastest growth rate mentioned, but in terms of progeny there is a great difference between a 16-min and a 20-min doubling time. In 24-h growth period at a 16-min doubling time there would be 2.6×10^5 times more organisms than at a 20-min doubling time).

E. coli must therefore be capable of contending with five major situations (Koch, 1971, 1976, 1985; Koch & Schaechter, 1985; Savegeau, 1974, 1983). They are: (1) chronic starvation, where nutrition is low, but continuing; (2) temporary superabundance, where the environment teems with adequate amounts and kinds of nutrients; (3) temporary total starvation; (4) getting from one mammalian intestine to another; and (5) enduring the transition between the other states. Only the first two have been mentioned above.

Looking at its metabolic and physiological accomplishments, one can see that it has adapted well to these several situations, all of which place severe, sometimes conflicting, restraints on the organisms. E. coli is parsimonious in terms of not carrying metabolic features that other organisms have that are apparently superfluous for this niche. It does not sporulate as do Gram-positive bacilli; it does not have the wide spectrum of metabolic capabilities as is found among the pseudomonads. It generally does not have the highly invasive characteristics of many pathogens. It is not too well-equipped to survive water stress. Its main strategy is simply to survive to outgrow its competitors in a limited set of circumstances and in a limited set of habitats.

There are other aspects to survival that will be mentioned, but not belaboured here, because they have been treated elsewhere (Koch & Schaechter, 1985). As a Gram-negative rod, the outer membrane protects it from bile and other toxic chemicals that are present in intestinal contents. It is adapted to the temperature of mammalian hosts; and the spectrum of organic chemicals that it can deal with are only those likely to be found in the intestine. In particular, these include several saccharides. Of these, the microbiologist and molecular biologists have made the most study and exploitation of its ability to metabolize lactose. But probably of equal importance is that it can grow very effectively on short-chain peptides; so while not proteolytic, it is very effective at scavenging short-chain polypeptides that result from proteolysis due to host enzymes and those of other proteolytic intestinal inhabitants. All in all, it is clear that the organism has survival skills for the special niche that it occupies.

I will turn now to the epidemiological facts concerning Escherichia coli (see Cooke, 1974; Draser & Hill, 1974; Skinner & Carr, 1974; Clarke & Bauchop, 1977; Linton, 1982; Simon & Gorbach, 1981). A newborn animal has a sterile intestine which rapidly becomes colonised, and a succession of organisms become dominant (see Linton, 1982; see p. 125-126 of Lynch & Poole, 1979; Smith & Crabb, 1961) in different regions of the gut. The gut finally attains a stable, but diverse, ecosystem in which the flora changes little thereafter. Of course, floral abundance in the gastroenteric tract of the non-ruminant is greatest in the colon, but organisms elsewhere in the mouth and small bowel are important to the host and determine the character of the bulk colon flora. Temporary displacements of the flora occur. For example, when a new strain of E. coli (usually one that produces enterotoxin) enters and succeeds in growing in the gut, there can be a disease-state produced where E. coli can become the dominant member of the population for a short period of time (as in Montezuma's revenge), but then it returns to the original minority status.

Under good nutritional and hygenic conditions, humans have 10^5 to 10^8 coliforms per gram of faeces. But these are central values and the fluctuations are extreme. In fact, the range extends to less than 10^2 organisms per gram of faecal contents in health to more than 10^{11} in special disease states; note that a packed cell mass would be about 10^{12}.

The coliform count of faecal or large intestine samples of carnivores and omnivores is usually larger than those of herbivores (Lynch & Poole, 1979). During babyhood, the coliform content, for a short period, can be much higher than the range when the mammal becomes an adult. Non-human animals have higher coliform counts than do humans. Even under quite constant and "clean" conditions, measurement of the faecal contents of two pigs sampled periodically over a period of time included a maximum 10^9 and 10^{10} coliforms per g. (Linton et al., 1978). Adults living in developing countries usually have considerably higher counts than do adult inhabitants of the "first-world". A good bit of this higher count, of course, is because of the larger input of coliforms into the host organism. We adult humans living in a civilized, hygienic world - a category into which probably most the readers of this article fit - sometimes have none and most often fewer than 10^7 coliforms per g of faeces (Finegold & Sutter, 1978). Civilized modern man probably is not of significance to E. coli, except as he engages in microbial and molecular genetics. In its captive state it has evolved much more in the last 30 years than in the previous 2 x 10^8 years of freedom!

THE BASIS OF TAXONOMIC LUMPING AND SPLITTING

A (presumptive) coliform count includes species other than E. coli, but what these organisms have in common is an ability to occupy very similar niches. Usually the count is based on the characteristic of being Gram-negative, facultatively anaerobic, non-spore-forming, rod-shaped, oxidase negative, fermenting lactose with the formation of acid and gas, bile- and detergent-resistant, peptide-consuming, and able to grow at mammalian temperatures. Growth at a higher temperature (45.5℃) is taken in the USA to exclude non-faecal organisms (see Goldreiche, 1976). The faecal coliforms are Gram-negative non-sporulating rods and are included in such genera as Escherichia, Citrobacter, and Klebsiella (Enterobacter) (see Bonde, 1966). However, members of genera life Shigella and Salmonella really occupy very similar niches, differing in failing to metabolize lactose.

The species other than E. coli that have been included in the coliforms or faecal coliforms have many metabolic and morphological properties in common. Is this because they have been "over classified" by the taxonomists? From the present point of view the differences are slight, although from a medical point of view some of these differences are important. In terms of their behaviour in a host where disease is not produced, there is little change in their ecology. Of course, we can tell the difference between cultures very easily based on detailed metabolic and serological properties of the Enterobacteriaceae (Edwards & Ewing, 1972; Ørskov, 1981). The metabolic reactions include positive methyl red, negative Voges-Proskauer, gas from D-glucose, growth on mannitol, no growth on citrate or malonate, no H_2S-production urease, or phenylalanine deaminase. Even within organisms classified as E. coli there are different subgroups based on metabolic criteria, and we can go further in subdivision by using antibiotic, phage colicin typing, enzyme mobility in gel electrophoresis, DNA hybridization, restriction length polymorphism, and the serotypes of surface antigens. One can further recognize the diversity by considering the degree of polymorphism in isozymes of various enzymes isolated from the organism. Again, there are a large number of types (Milkman, 1973; Levin, 1981; Whittam et al., 1983). With all the variation detected this way, the conclusion can be drawn that there is a very large number of kinds of organisms occupying this "coliform" niche.

Diversity Within a Bacterial Species

All of this points out that bacteria with many taxonomic characters in common can still be highly diverse. It is important to understand that taxonomy of prokaryotes can almost never produce a clear-cut definition because any organism that has even a single gene change from its parent represents a whole new clone, and in some important sense, a whole new species of organism. This circumstance is completely different for obligate sexual organisms that must form an interbreeding population so that any not too deleterious mutation will have a chance to circulate within the entire group. Consequently the characteristic is associated with the group but need not be expressed in any one individual. With justification, we could further lump or split the species of Enterobacteriaceae and, except in the case of extreme splitting, there would be a great deal of polymorphism. But the big question is: Why is there so much diversity and polymorphism within the coliforms? Formulated in another way, the major subject that we must try to understand is why there can be so many quite similar, but still different organisms occupying seemingly the same niche.

I have collected seven different explanations for this kind of diversity. Seven is a larger number than is usually considered by ecologists (Pianka, 1983; Emlen, 1984). The first and major one is that different organisms truly occupy different niches, even though the biologist may not have understood what these differences are nor why they are important. The second explanation is that there is heterogeneity in the quality of life habitat available to the organism. This "patchiness" in space permits different genotypes to prosper in different locations. The third is that there is a temporal heterogeneity of conditions that sometimes favours one type of organism and at other times another, in certain cases conditions are such that when averaged over seasons or long times there can be stable coexistence that will persist indefinitely (Koch, 1974). The fourth possibility is that the differences being considered are indeed truly neutral and in fact either type is exactly as fit as the other. The fifth is that actually one organism is attaining dominance, but that the replacement process is sufficiently slow and was started recently enough that the human observers have caught the process only in transition. The sixth explanation is that there is actually a positive selection against sameness. Density-dependent phenomena such as these are beginning to be understood for higher organisms which possess sensory and mental capabilities that allow them to identify conspecifics and individuals. But density-dependent phenomena can also occur in situations with prokaryotic organisms with their very limited sensory and memory capacities because the host is not so limited. For example, if the host makes IgA antibodies against an intestinal organism, this would select against the previously abundant form and leave a void that could be filled by any variant not binding that antibody. In such cases, groups of organisms with diverse serotypes would be favoured.

The Resident Strain Phenomenon and the Seventh Explanation

There is clear evidence for "sameness" in the resident strain phenomena discovered thirty-five years ago by Sears (Sears et al., 1950, 1956; Sears and Brownlee, 1952; Hartley et al., 1977; see Hartl & Dykhuizen, 1984). His group observed that E. coli isolated from the gut of individual humans could have the same serotype over a time interval as long as several years. These serotypes would eventually be replaced and then the new serotype could also remain as the dominant "resident" for a long period of time. However, even during such intervals there occasionally could be transient organisms with different serotypes that would appear in the colon ecosystem, persist for a while and later disappear. The interpretation was that somehow the resident strain had some special advantage during its tenure whereas the transients could not establish a "foothold" in the gut ecosystem. Later studies have extended the observations on this basic phenomenon; the concept certainly has held up for gut ecosystems of adult humans in developed countries, although the coliform flora at any one time appears to be richer than Sears envisaged.

Due to the work of Rolf Freter (Freter, 1980; Freter et al., 1983) we can now partially understand the maintenance of the resident strain by the favourable positioning of the resident organism in the surface layers of the colon. His group demonstrated in both suitable gnotobiotic mice and in a continuous-flow apparatus that the part of the flora adhering to the gut wall (or to other bacteria adhering to the wall) served as a continuing source of bacteria for the lumen. He was able to show mathematically that coexistence of strains occurs such that the majority strain is the first strain that occupied the sites on the wall surfaces. These ideas are clearly pertinent, but the model as developed by Freter so far does not account for the extreme dominance of one serotype and its rare, essentially total, replacement.

There is reason to believe that the resident-strain idea also applied to animals in a somewhat different way. The best studied case is that reported by Linton et al., (1978) in which two pigs, after having been weaned from their mothers, were taken to separate pens in different buildings, fed by different handlers, and samples of their flora analysed many times throughout the pigs' lives for the serotype of their coliforms. At each sampling time the serotype of a hundred independent coliforms was determined. The pigs were in good health and, save for a very small amount of tetracycline that had been given their mothers, one of them was never exposed to antibiotics. The other was given an experimental treatment with tetracycline. The experimental results concerning antibiotic resistance are irrelevant for our point of view, although the study of the spectrum of resistance types provides additional markers to the serotypes.

Fig. 3. Appearance and disappearance of serotypes of *Escherichia coli* in the faecal contents of a single pig. With permission from Linton et al., (1978).

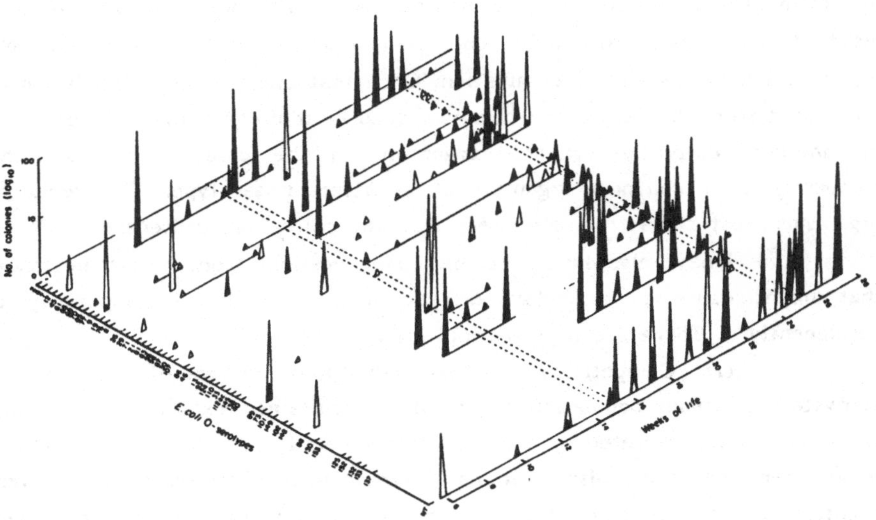

The overall finding (see Fig. 3) was that many different serotypes appear very occasionally and go away quickly as do the transients observed in the intestines of man. A particular serotype would rise from 0 out of 100 to majority status and then fall within a few days back to 0. The especially interesting finding was that many of these serotypes rose and fell on a number of separate occasions. So, for these pigs, there were a number of serotypes that could be considered as "alternate co-resident" strains. However, each serotype spent a majority of its time in seclusion or at least being present in the faeces in such small numbers so as not to be detected at all in an analysis of 100 independent isolates. Thus, being "in residence" in the pig's colon is not as continuous as it had seemed to be in a first-world human intestine.

Explanations for Resident and Coresident Strains

How are we to explain this episodic growth? The rapid rises imply that the large bowel is a good place to live and grow in all respects for E. coli (at least under special circumstances and location). The temperature is optimum, the water is sufficient, and while on the average the colon flora must have 12 h doubling time (Gibbons & Kapsimalis, 1967; Gorbach & Leviton, 1979), limited no doubt by the entry rate of resources, there can be times when growth conditions are much more favourable than that and the doubling times may locally and temporarily be as fast as the maximum rate at which the organism can grow. This would lead to a rapid burst of growth that then elicits counter-measures against the now abundant organism leading to virtual elimination. Once that clone of organism has been eliminated from the colon, an ecological vacuum is created that will be filled by another, currently available organism. In the case of the pig this frequently is a coliform organism of a different serotype. The replacing organisms must come from elsewhere than the colon bulk contents.

I have elsewhere categorized the possible types of circumstances that might cause a particular serotype to be maintained temporarily or permanently. There are nine possibilities :

(1) Competition/symbiosis with other members of the intestinal ecosystem. A steady population might not be maintained because some vitamin or growth factor excreted by the coliforms generally stimulates the growth of some other kind of organism that may increase in numbers (possibly only very slightly) and temporarily decreases the availability of nutrients otherwise consumed by the coliforms. When such nutrients are exploited in this way, the

coliforms may die out for the reason similar to that for which algal blooms in freshwater lakes die out. Subsequently, these other organisms or classes of organisms, for lack of the coliform-produced growth factor, become reduced in numbers, re-establishing the habitat for coliforms. Any of the many serotypes of coliform entering the colon may be the opportunist that succeeds in growing very rapidly.

(2) Failure of coliforms to respond sufficiently rapidly to "shift-down" conditions. During rapid growth, the organisms in vivo do the same thing they are known to do in vitro; i.e. repress the biosynthetic pathways for amino acids,. vitamins, purines, pyrimidines, and production of chelators for iron, etc. When all those resources become depleted almost simultaneously, the coliforms have to grow much more slowly based on other more intractable or slowly-utilized resources. Under these unfavourable conditions, there may be a shift-down and consequent slowing of growth to such a degree that growth slows so far that the organisms are washed out by faecal flow or eliminated by specific and non-specific mechanisms which were operating all along.

(3) Response to host immune mechanisms. Production by the host of antibodies of the IgA type present on cells and secreted into the colon may take place as a secondary (amnestic) response to the rapid growth of a serotype and may serve to end its bloom.

(4) Development of allelopathic response(s) by other organisms in the gut. A whole host of organic compounds and antibiotics are secreted in the intestine. These include microcins and high levels of volatile fatty acids. The temporary production or extra high production of such compounds may temporarily eliminate a subclass or all of the coliform population. The selectivity and generality of the antibiosis would vary according to the chemical species. This model is the weakest of nine possibilities because I know of no mechanism to lead to the reduction in antibiotic production except in conjunction with possibly (1).

(5) Colicin production. Certain strains of E. coli produce colicins - special proteins that are lethal to other sensitive strains of coliforms. This could lead to the replacement of a dominant serotype if the producing strain was of a different serotype. The latter in turn could be replaced by yet another type of colicin producer.

(6) Bacteriophage production and killing of the majority of the organisms in a bloom. Only when a sensitive population of E. coli becomes sufficiently large, can the lytic growth of the bacterial viruses that might be

present at low levels at all times become explosively autocatalytic and act to destroy the bulk of the coliform population (Levin & Lenski, 1983).

(7) The successive replacement of strains with different metabolic "specialities". This model could also lead to the maintenance of resident strains or to the alternation between serotypes, depending on circumstances. It would work as follows: During times of almost continuous supply, of a nutrient such as lactose in the gut of a nursing baby, consumption by inducible strains would lead to a chronic limitation of the lactose. Although a considerable amount entered the system, the free concentration remains low. Mutants or other strains constitutive for the production of lac operon products would be selected and replace the dominant type. This replacement, if by a mutant of the original strain, would maintain the serotype. That strain, secondarily, might be replaced by a strain constitutive for some other nutrient (say, when the baby is weaned). Again, maintenance of the serotype would be most likely if the variant arose by mutation, but less likely if many strains, even if at very low incidence, were present. Gradually, as other nutritional factors become important, the genetic state of regulation for an earlier important selective factor would replace the original wild-type inducible or repressible state by mutation and selection. Overall, this could lead to the long-term retention of a serotype because one at a time, certain genetic characters, unrelated to serotype, come under intense selection. Replacement of serotype would only occur when the new transient serotype happened to have exactly the right constellation of mutations of regulatory genes. In the absence of being bombarded with many strains, the replacement would arise mutationally. In the pig the serotype would change frequently because there are a plethora of strains available.

(8) Cyclic replacement. Paquin & Adams (1983) has recently shown that in chemostat culture periodic selection leads to replacement by strains that are favoured by their immediate predecessor, but are at a disadvantage relative to an earlier stain. The basis of these selective advantages are not totally clear, but at least this partially has to do with rendering metabolites inaccessible to the other strain. In the gut ecosystem this could lead to a cycle of strain substitutions. Such a cycle of replacements would maintain the serotype and would also allow the replacement of the dominant serotype by other serotypes, if many strains were available.

(9) Immigration from the ileum. Although the vast bulk of the coliforms reside in the colon, I have argued elsewhere (Koch, 1985) that the

regulatory mechanisms present in these bacteria clearly evolved to serve a role in the small intestine where such nutrients are available in reasonable concentrations. One can argue that Escherichia coli should have its name changed to Escherichia ilei to express the notion that the ileum is the site of growth of the "founders" of the population in the lower bowel. Although there are few coliforms per ml of ileal fluid (Draser & Hill, 1974; Ørskov, 1981), they could be growing at their maximal rate; however, because the dilution rate (flow rate/culture volume) is so fast, such a population will be washed out if not replenished. The same condition would also apply to the first part of the ascending colon. There, rapid flow of fluid, incompletely mixed, leads to "plug-flow growth". This mode of growth could lead to all of the aspects of coliform growth described above. First, the wild fluctuations in total coliform level could result from random sloughing of patches of organism from the ileal surface. These sloughed cells would grow rapidly in the ileum and first part of the large intestine, but would grow very slowly or not at all in the distal part of the colon. During periods where sloughing from the ileal surface happened not to occur, the level of coliforms would decrease due to the washout of the lumen part of the tract. Of course, the resources that otherwise the coliforms would have consumed would then be exploited by other organisms.

CONCLUSIONS

The properties of the enteric microorganism are the special result of the vast number of organism generations that have adapted them to a very specialised niche in a specialised habitat. This has made for a diversity at the biochemical and microbial physiological level. In previous papers I have argued that even more important than the cumulative number of organisms is the number of recolonising events leading to cultures that may fix certain mutants in the population. This has clearly led to groups of similar organisms that are maintained by the options available given their basic strategy.

Work in my laboratory is supported by grant GM34222 from the United States Department of Health Services. The ideas expressed here were developed over a number of years but were brought to the present state of semi-fruition under the aegis of Marilyn Bowie, Moe Goldner and George Hegeman.

REFERENCES

Bonde, G.J. (1966) Bacterial indicators of water pollution. In: Advances in Aquatic Microbiology, vol. 1, Ed, M.R. Droop & H.W. Jannasch,

pp. 273-364. Academic Press, London.
Clarke, R.T.J. & Bauchop, T. (1977) Microbial Ecology of the Gut. Academic Press, London.
Cooke, E.M. (1974) Escherichia coli and Man. Churchill Livingstone, Edinburgh.
Draser, B.S. & Hill, M.J. (1974) Human Intestinal Flora. Academic Press, London.
Edwards, R.R. & Ewing, W.H. (1972) Identification of Enterobacteriaceae. 3rd Edn. Burgess Publication Co. Minneapolis.
Emlen, J. (1984) Population Biology : The Co-evolution of Population Dynamics and Behaviour. Macmillan, New York.
Finegold, S.M. & Sutter, V.L. (1978) Faecal flora in different populations with special reference to diet. Amer. J. Clin. Med. 31, S116-22.
Freter, R. (1980) Prospects for preventing the association of harmful bacteria with host mucosal surfaces. In: Bacterial Adherence. Ed. E.H. Beachey. pp. 439-458. Chapman and Hall, London.
Freter, R. et al. (1983) Continuous-flow cultures as in vitro models of the ecology of large intestinal flora. Infect. & Immunol., 39, 666-75.
Gibbons, R.J. & Kapsimalis, B. (1967) Estimates of the overall role of growth of intestinal microflora of hamsters, guinea pigs and mice. J. Bacteriol., 93, 510-12.
Goldreiche, E.E. (1976) CRC critical reviews of environmental microbiology. Microbiology, 6, 349-69.
Gorbach, S.L. & Leviton, R. (1979) Intestinal flora in health and gastrointestinal diseases. In: Progress in Gastrointerology, Vol. 2, Ed. G.B.J. Glass, pp. 478-84. Grune & Stratton, New York.
Hartl, D.L. & Dykhuizen, D.E. (1984) The population genetics of Escherichia coli. Ann. Rev. Genet., 18, 31-68.
Hartley, C.L. et al. (1977) Escherichia coli in the faecal flora of man. J. App. Bact., 43, 261-69.
Koch, A.L. (1971) The adaptive responses of Escherichia coli to a feast and famine existence. Adv. Microb. Phys., 6, 147-217.
Koch, A.L. (1974) Coexistence resulting from an alternation of density dependent and density independent growth. J. Theor. Biol., 44, 381-95.
Koch, A.L. (1976) How bacteria face depression, recession and derepression. Pers. Biol. & Med., 20, 44-63.
Koch, A.L. (1984) Evolution vs. the number of gene copies per primitive cell. J. Mol. Evol., 20, 71-6.
Koch, A.L. (1985) The co-evolution of information and energy transfer with the origin of cells. J. Mol. Evol., 21, 270-7.
Koch, A.L. (1985) The macroeconomics of bacterial growth. In: Bacteria in Their Natural Environments. Eds. M.M. Fletcher & G.D. Floodgate, pp. 1-42. Amer. Soc. Microbiol.
Koch, A.L. & Schaechter, M. (1985) The world and ways of E. coli. In: Industrial Microorganisms, Vol. 1, Ed. A. Demain, pp. 1-25. Addison/Westley, Reading, Mass.
Levin, B. (1981) Periodic selection, infectious gene exchange and genetic structure of E. coli populations. Genetics, 99, 1-23.
Levin, B.R. & Lenski, R.E. (1983) Co-evolution in bacteria and their viruses and plasmids. In: Co-evolution, Eds. D.J. Futuyma & M. Slatkin, pp. 99-127. Sinauer Assoc., Inc. Sunderland, Ma.
Linton, A.H. (1982) Microbes in Man and Animals. Wiley, London.
Linton, A.H. et al., (1978) Fluctuations in Escherichia coli O-serotypes in pigs throughout life in the present and absence of antibiotic treatment. J. App. Bact., 44, 285-8.
Lynch, J.M. & Poole, N.J. (1979) Microbial Ecology : A Conceptual

Approach. Blackwell Scientific Publications, Oxford.

Milkman, R. (1973) Electrophoretic variation in E. coli from natural sources. Science, 182, 1024-6.

Ørskov, F. (1981) Escherichia coli. In: The Prokaryotes, Vol. 2, Eds. M.P. Starr, H. Stolp, H.G. Truper, A. Balows & H.G. Schlegel, pp. 1128-34. Springer-Verlag, Berlin.

Pianka, E.R. (1983) Evolutionary Ecology. 3rd Edn. Harper and Row, New York.

Paquin, C.E. & Adams, J. (1983) Relative fitness can decrease in evolving asexual populations of S. cerevisiae. Nature, 306, 368-71.

Savegeau, M. (1974) Genetic regulatory mechanisms and ecological niche of Escherichia coli. Proc. Nat. Acad. Sci., USA, 71, 2453-55.

Savegeau, M. (1983) Escherichia coli habitats, cell types and molecular mechanisms of gene control. Am. Nat., 122, 732-44.

Sears, H.J. et al. (1950) Persistence of individual strains of Escherichia coli in the intestinal tract of man. J. Bact., 59, 293-301.

Sears, H.J. & Brownlee, I. (1952) Further observations on the persistence of individual strains of Escherichia coli in the intestinal tract of man. J. Bact., 63, 47-57.

Sears, H.J. et al. (1956) Persistence of individual strains of Escherichia coli in man and dog under varying conditions. J. Bact., 71, 370-372.

Simon, G.L. & Gorbach, S.L. (1981) Intestinal flora in health and disease. In: Physiology of the Gastrointestinal tract, Ed. in chief L.R. Johnson, pp. 1361-80. Raven Press, New York.

Skinner, F.A. & Carr, J.G. (1974) The normal flora. Society for Applied Bacteriology, Symposium No. 3. Academic Press, London.

Smith, H.W. & Crabb, W.E. (1961) The faecal bacterial flora of animals and man: its development in the young. J. Path. Bact., 82, 53-66.

Whitham, T.S. et al. (1983) Multilocus structure in natural populations of Escherichia coli. Proc. Natl. Acad. Sci. USA, 80, 1751-5.

Woese, C.R. et al. (1985) The phylogeny of purple bacteria: the gamma subdivision. Syst. App. Microbiol., 6, 25-33.

EVOLUTIONARY PHYSIOLOGICAL ECOLOGY OF PLANTS

J.P. Grime
R. Hunt
W. J. Krzanowski

INTRODUCTION

A century after Darwin, the central unsolved problem of ecological research remains that of identifying the various selection processes which have determined the diversity of the planet's animals and plants (Hutchinson 1959), their current distribution and their roles within communities and ecosystems. This task consists essentially of a dissection of the 'struggle for existence' (Darwin 1859) by recognizing first, the circumstances in which particular forms of selection occur, second, the genetic traits upon which they operate and third, the design constraints (Grime 1965) which channel evolutionary responses to selection. In view of the widespread acceptance of Darwin's theory it seems not unreasonable to apply his methods in seeking solutions to these contemporary problems.

Darwin's method of research was to turn his extraordinarily active curiosity upon many aspects of the biology of the wide range of organisms which he encountered as naturalist, paleontologist and world-traveller. There can be little doubt that the theory of natural selection owes much to this rich comparative experience. It is perhaps ironic, therefore, to find that the broad comparative methods which led to Darwin's achievement remain controversial (Gould & Lewontin 1979, Clutton-Brock & Harvey 1979) when applied to contemporary problems in evolutionary ecology. Indeed, the strictures applied by some critics of the comparative approach (Harper 1982, Woolhouse 1981) lead to the suspicion that Darwin himself might face charges of 'facile guesswork' if he was to return to venture opinions in the current debate.

Elsewhere, Grime (1984, 1985) has attempted to explain how modern trends of specialization in biology have led to a decline in the use of Darwin's method. This paper will suggest that broadly-based comparative studies remain an essential component of evolutionary and ecological research. We shall contend, however, that full value will be drawn from this approach only where the conventional emphasis upon morphology, demography and reproductive schedules is extended to include physiological attributes of the organisms under study. Our examples refer exclusively to plants; the principles apply more widely.

INDIVIDUAL TRAITS

Several generations of plant ecologists have now sought to recognize patterns of ecological specialization by reference to variation in specific attributes of plants. Plant characteristics which have been found useful for this purpose include life-span and morphology (Ramensky 1938, Shepherd 1981), the position of perennating buds (Raunkiaer 1934), leaf form (Givnish 1982, Box 1981), shoot phenology (Al-Mufti et al. 1977), the size of seeds (Salisbury 1942, Baker 1972) and the persistence of seeds in the soil (Chippindale & Milton 1934, Roberts 1970, van der Valk & Davis 1976, Thompson & Grime 1979). Criteria based upon comparative study of plants in the laboratory may be added to this list; these include the antiherbivore defences of leaves (Grime et al., 1968, Reader & Southwood 1981, Coley 1983), relative growth rates of seedlings (Grime & Hunt 1975) and seed germination responses to chilling, scarification, temperature and irradiance (Kinzel 1920, Taylorson & Borthwick 1969, Gorski 1975, Grime et al. 1981).

Many opportunities remain to improve the predictive power of models of ecological specialization through incorporation of additional physiological attributes. Two recent examples will illustrate this point.

Nuclear DNA Content as an Index of Phenology and Climatic Response

It is established that the chromosome sizes and nuclear DNA contents of vascular plants in the tropics are consistently small in comparison with the wide variation encountered in temperate regions (Avdulov 1931, Stebbins 1956, Levin & Funderburg 1979, Jones & Brown 1976, Bennett & Smith 1976, Bennett 1976). In the temperate zone and particularly in Mediterranean climates, high nuclear DNA contents are characteristic of vernal geophytes and grasses. In Northern England a positive correlation has been established between genome size and the rates of leaf expansion and production of shoot biomass in the early spring and a mechanism has been proposed to explain variation in nuclear DNA amount as a product of climatic selection (Grime & Mowforth 1982, Grime 1983, Grime et al., 1985). As shown in Fig. 1, nuclear DNA contents may vary considerably within the same stand of vegetation and may be used as an index of temporal niche differentiation within the community.

Leaf and Root Plasticity as an Index of Resource 'Foraging' and Competitive Ability

Many vascular plants are capable of dramatic changes in morphology in response to variation in habitat factors such as temperature, irradiance, water supply and frequency of defoliation. It is generally assumed that this morphological plasticity makes an important contribution to the ability of certain common plants to exploit contrasted habitat conditions or to persist in vegetation experiencing fluctuations in physical conditions or vegetation management (Bradshaw 1965, McNaughton et al. 1974). Recently however an additional hypothesis has been put forward to explain the functional significance of morphological plasticity (Grime 1979). This theory proposes that environmentally induced plastic changes in morphology are an integral part of the mechanism of resource acquisition in some plants. Crick (1985) explains how many vascular plants although sessile 'have the capacity to produce new meristems throughout their life and exhibit a capacity for structural variance unequalled by most non-sessile animals' and Bookman & Mack (1982) point out that 'for plants, root system and leaf canopy development is analagous to foraging for resources in three-dimensional space.' In natural vegetation an individual plant encounters a complex and rapidly-changing patchiness in resource supply above and below ground and it is likely that genotypes exhibiting rapid plastic responses in root and shoot development will be at a selective advantage if the result of these responses is to concentrate leaves and roots in the resource-rich parts of the environment.

According to the hypothesis of Grime (1979), rapid morphological responses in the development of leaves and roots may be of selective advantage only for plants in productive habitats where the returns for 'foraging' are high because resource-rich patches are extensive and relatively long-lived. In circumstances where the productivity is low and resource availability is brief and unpredictable (e.g. during sun-flecks or nutrient pulses of very short duration) we may expect less plasticity. This is because conservation of captured resources is of prime importance in these stressed environments and liberal expenditure of energy and mineral nutrients in morphological change could result in a net loss of resources (Sibly & Grime, 1986).

Fig. 1. Nuclear DNA content and leaf extension rate in March/April among nineteen Cressbrookdale species. 1. Agrostis vinealis; 3. Avenula pratensis; 4. Briza media; 5. Dactylis glomerata; 6. Danthonia decumbens; 7. Festuca ovina; 8. Festuca rubra; 9. Koeleria macrantha; 10. Carex caryophyllea; 11. Carex flacca; 12. Carex panicea; 13. Carex pulicaris; 14. Anemone nemorosa; 15. Leontodon hispidus; 16. Plantago lanceolata; 17. Ranunculus bulbosus; 18. Sanguisorba minor; 19. Succisa pratensis; 20. Viola riviniana. 95% confidence limits are indicated; r = 0.606.

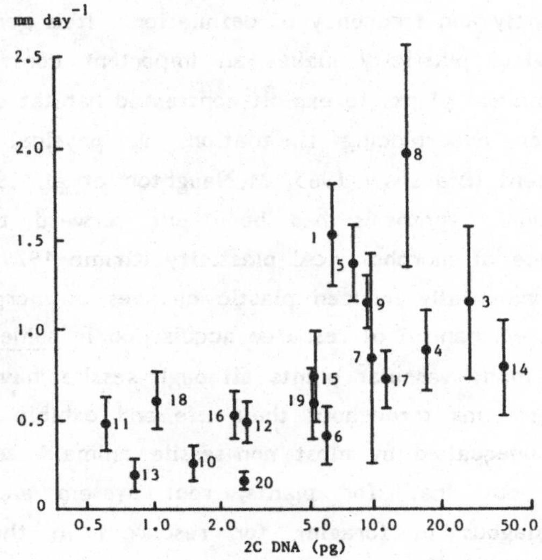

Fig. 2. Leaf and root plasticity in grasses: (a) Shoots of Agrostis stolonifera in a patchy light environment; (b) Roots of Holcus lanatus in a patchy nutrient environment. (2C DNA, pg)

Recently experiments have been conducted to compare the ability of plants of contrasted ecology to concentrate their leaves and roots in the resource-rich sectors of partitioned root or shoot environments (Fig. 2). The results are consistent with the theoretical predictions and will be reported in detail elsewhere.

SETS OF TRAITS

As we have seen in the preceding section, some individual plant attributes have considerable predictive value. This has led some ecologists and in particular Noble & Slatyer (1979) to propose a system of 'vital attributes' upon which to base predictions of plant succession and vegetation responses to perturbation. Whilst there can be little doubt of the value of this approach as a management tool in specific systems, it does not provide a basis for study of the processes involved in evolutionary and ecological specialization. There are two main difficulties associated with the use of vital attributes in ecological analysis:

(1) The same attribute may have an entirely different significance in two different organisms, e.g. a high relative growth rate may allow rapid completion of an ephemeral life history in a temporary habitat (e.g. Papaver rhoeas) or facilitate long-term occupation of a stable productive habitat (e.g. Reynoutria japonica).

(2) It is rarely profitable to examine variation in a single attribute without reference to other characteristics of the organisms under study. Ecological specialization is usually associated with correlated changes in a set of traits (Hutchinson 1951, MacArthur & Wilson 1967, Pianka 1970).

THE C-S-R MODEL

Comparisons of plants drawn from contrasted habitats within local floras reveal numerous recurring sets of traits which can be related to functional specializations. These may be associated with a very specific activity, e.g. insect pollination in a distinct group of plants such as orchids. At the other extreme however we may recognize a suite of attributes which engages most of the fundamental activities of the plant, recurs throughout the Plant Kingdom and is represented in all of the world's major biomes. It is of course in this latter case that we are concerned with the attempt to make a primary dissection of Darwin's Struggle for Existence. In botanical terms success in this enterprise will be measurable in terms of the ability to classify plants into the minimal number of functional types (primary strategies) necessary to explain the structure and dynamics of vegetation.

The most widely known theory of primary strategies is that of r- and K-selection (MacArthur & Wilson 1967). Despite its many attractive features, this theory has two fundamental weaknesses (Greenslade 1972a,b, Wilbur et al. 1974, Gill 1978, Grime 1979) which can be summarized as follows:

(1) The concept fails to recognize a third primary strategy (stress-tolerance) of widespread occurrence in stable unproductive habitats or niches.

(2) The theory does not take into consideration the fact that organisms may exhibit quite distinct strategies as juveniles and adult organisms.

The C-S-R model of primary strategies attempts to rectify these omissions and rests upon the assertion (Ramensky 1938, Grime 1974) that the primary mechanism controlling the nature and distribution of plant populations, species and communities arises from three selection processes which operate in the present and have also exerted a dominant influence in the evolution of plants. These selection processes may be defined as stress, consisting of the external constraints on dry-matter production, and disturbance, identified as the destruction of biomass. At high intensities both stress and disturbance select for particular types of life history and physiology. Where the intensities of stress and disturbance are low, rapid rates of resource capture and growth are possible and a third selection process (competition for light, water, mineral nutrients and space) becomes of overriding importance and selects for a quite different set of plant characteristics. In addition to the strategies associated with high intensities of either stress (stress-tolerators), disturbance (ruderals) or competition for resources (competitors), there are others which exploit the various intermediate conditions, corresponding to particular equilibria between the three selection forces; these may be described by means of a triangular model. A more detailed account of these concepts is available in Grime (1979) which also contains a description of five primary regenerative (juvenile) strategies of widespread occurrence in plants and their inter-relationships with C, S and R. Table 1 presents the sets of attributes, many of them physiological, which have been associated with the three primary strategies. Information consistent with the C-S-R model is now available for herbaceous plants (Grime 1974, Chapin 1980, Leps et al. 1982), bryophytes (Furness 1979), Algae (Raven 1981, Shepherd 1981, Coesel 1982, Dring 1982), fungi (Pugh

Table 1. Some characteristics of competitive, stress-tolerant and ruderal plants.

		Competitive	Stress-tolerant	Ruderal
(i)	Morphology			
1	Life forms	Herbs, shrubs and trees	Lichens, bryophytes, herbs, shrubs and trees	Herbs, bryophytes
2	Morphology of shoot	High dense canopy of leaves. Extensive lateral spread	Extremely wide range of growth forms	Small stature, limited lateral spread
3	Leaf form	Robust, often mesomorphic	Often small or leathery, or needle-like	Various, often mesomorphic
4	Canopy structure	Rapidly elevating monolayer	Often multi-layered. If monolayer not rapidly elevating	Various
(ii)	Life-history			
5	Longevity of established phase	Long or relatively short	Long or very long	Very short
6	Longevity of leaves and roots	Relatively short	Long	Short
7	Leaf phenology	Well-defined peaks of leaf production coinciding with periods of maximum potential productivity	Evergreens, with various patterns of leaf production	Short phase of leaf production in period of high potential productivity
8	Phenology of flowering	Flowers produced after (or, more rarely, before) periods of maximum potential productivity	No general relationship between time of flowering	Flowers produced early in the life-history
9	Frequency of flowering	Established plants usually flower each year	Intermittent flowering over a long life history	High frequency of flowering
10	Proportion of annual production devoted to seeds	Small	Small	Large
11	Perennation	Dormant buds and seeds	Stress-tolerant leaves and roots	Dormant seeds
12	Regenerative* strategies	V, S, W, B_S	V, W, B_{SD}	S, W, B_S
(iii)	Physiology			
13	Maximum potential relative growth rate	High	Low	High
14	Response to stress	Rapid morphogenetic responses (root:shoot ratio, leaf area, root surface area) maximizing vegetative growth	Morphogenetic responses slow and small in magnitude	Rapid curtailment of vegetative growth, diversion of resources into flowering
15	Photosynthesis and uptake of mineral nutrients	Strongly seasonal, coinciding with long continuous period of vegetative growth	Opportunistic, often uncoupled from vegetative growth	Opportunistic, coinciding with vegetative growth
16	Acclimation of photosynthesis, mineral nutrition and tissue hardiness to seasonal change in temperature, light and moisture supply	Weakly developed	Strongly developed	Weakly developed
17	Storage of photosynthate mineral nutrients	Most photosynthate and mineral nutrients are rapidly incorporated into vegetative structure but a proportion is stored and forms the capital for expansion of growth in the following growing season	Storage systems of leaves, stems and/or roots	Confined to seeds
(iv)	Miscellaneous			
18	Litter	Copious, often persistent	Sparse, sometimes persistent	Sparse, not usually persistent
19	Palatability to unspecialized herbivores	Various	Low	Various, often high
20	Genome size	Usually small	Various	Small or very small

*Key to regenerative strategies: V, vegetative expansion; S, seasonal regeneration in vegetation gaps; W, numerous small wind-dispersed seeds or spores; B_S, persistent seed bank; B_{SD}, persistent seedling bank.

1980, Cooke & Rayner 1984) and corals (Rosen 1981).

A TEST OF STRATEGY CONCEPTS

One method of testing strategy hypotheses, such as the C-S-R model, is to use multivariate analysis to examine the characteristics of the species present in local floras to seek evidence for or against the co-occurrence of plant attributes in the sets predicted by the strategy concepts.

A variety of ways is available for exploring axes or groupings within multivariate data sets. Principal components analysis, for example, identifies axes of specialization to which the different attributes variously contribute, both in kind and in degree. But perhaps the simplest foundation for an analysis is the assumption that the values of certain attributes for a particular group of species have evolved through simultaneous natural selection in a composite response to various external environmental pressures. Further, they may well have evolved in such a way that coherent groupings of species can now be identified, each with characteristic values or ranges of values in each of the attributes. This being so, a cluster analysis is indicated.

Two variants of cluster analysis exist. The first employs an hierarchical method in which each species belongs initially to its own separate class. The two classes with the highest similarity are then merged and all other classes are redefined according to an agreed criterion. This process is then repeated until all species belong to a single class, intermediate stages in this process providing the information sought. Since it is less likely that co-evolution of plant attributes will have occurred in response to fine

Table 2 (opposite). The Unit of Comparative Plant Ecology (UCPE) database used in the multivariate analysis. Attributes 1-24 are derived from the published literature, from laboratory screening procedures and our own field observations. Attributes 25-30 rely upon our own field sampling. The 'stress' and 'disturbance' indices (26, 27) differ from the remaining environmental attributes (25, 28, 29 and 30) in that they are not based upon environmental measurements. They are indirect assessments of the intensities of stress and disturbance experienced by the species in their natural habitats. These assessments are based upon the frequency of occurrence of selected plant attributes in the vegetation samples found to contain the species. A description of the procedure used to derive these stress and disturbance co-ordinates is contained in Grime (1984). Attributes 26 and 27 are not used in clustering.

(a) Vegetative attributes

1	DNA 2C DNA content	1	<2.0 pg
		2	2.0-3.9 pg
		3	4.0-9.9 pg
		4	10.0-19.9 pg
		5	20.0 pg and above
2	HEIGHT Canopy height	1	<250 mm
		2	260-500 mm
		3	510-750 mm
		4	760-1000 mm
		5	>1 m
3	SPREAD Lateral spread	1	Therophytes
		2	Compact perennials <100 mm
		3	Perennials 110-250 mm
		4	Perennials 260-1000 mm
		5	Perennials >1 m
4	FTIME Flowering time	1	Jan., Feb. or March
		2	April
		3	May
		4	June
		5	July onwards
5	FDUR Flowering duration	1	Up to 1 month
		2	Up to 2 months
		3	Up to 3 months
		4	Up to 4 months
		5	>4 months
6	LIFEFORM Life form (Raunkiaer)	1	Therophyte
		2	Geophyte
		3	Hemicryptophyte
		4	Chamaephyte
		5	Phanerophyte
7	RGR Relative growth rate (seedling phase)	1	<0.5 $week^{-1}$
		2	0.5-0.9 $week^{-1}$
		3	1.0-1.4 $week^{-1}$
		4	1.5-1.9 $week^{-1}$
		5	2.0 $week^{-1}$ and above
8	LSHAPE Leaf blade shape (length/breadth)	1	50 and above
		2	25-49
		3	5-24
		4	2.5-4.9
		5	<2.5
9	LSIZE Leaf blade size (length x breadth)	1	<400 mm^2
		2	400-999 mm^2
		3	1000-2499 mm^2
		4	2500-4999 mm^2
		5	5000 mm^2 and above
10	LIFEHIST Life history	1	Ephemeral
		2	-
		3	Monocarpic, not perennial
		4	-
		5	Polycarpic
11	LPHENOL Leaf phenology	1	Leaves in summer only
		2	-
		3	Leaves mainly in summer
		4	-
		5	± evergreen
12	STOR Storage organs	1	In seeds only
		2	-
		3	Throughout the plant
		4	-
		5	In specialised organs
13	LTEXT Leaf texture	1	Succulent
		2	-
		3	Mesomorphic, not succulent
		4	-
		5	Hard, wiry or tough
14	LHAIR Leaf hairiness	1	Not hairy
		2	-
		3	Somewhat hairy
		4	-
		5	Very hairy
15	BREED Breeding system	1	Outbreeding
		2	Mainly outbreeding
		3	Mainly inbreeding
		4	Inbreeding
		5	Apomictic

(b) Reproductive attributes

16	DSPACE Dispersal in space	1	<1 mm, wind-dispersed
		2	-
		3	Wind, water or animals
		4	-
		5	No such dispersal
17	SWT Seed (disseminule) fresh weight	1	<0.1 mg
		2	0.2-0.5 mg
		3	0.6-1.0 mg
		4	1.1-5.0 mg
		5	>5.0 mg
18	GINIT Initial germinability	1	>75%
		2	50-74%
		3	25-49%
		4	<25%
		5	Nil
19	GRANGE Germination range at constant temp.	1	No germination
		2	4-12 °C
		3	13-20 °C
		4	21-28 °C
		5	>29 °C
20	GDARK Germination in darkness	1	>90%
		2	50-89%
		3	10-49%
		4	<10%
		5	Nil
21	GMODE Germination mode at constant temp.	1	<15 °C
		2	15-20 °C
		3	21-25 °C
		4	25-30 °C
		5	>30 °C
22	STEXT Seed texture	1	Smooth
		2	-
		3	Rugose, tuberculate etc.
		4	-
		5	Hairy or spiny
23	SSHAPE Seed shape (length/breadth)	1	<1.5
		2	-
		3	1.5-2.5
		4	-
		5	>2.5
24	DTIME Dispersal in time	1	Seed bank nil or type I
		2	-
		3	Seed bank type II or III
		4	-
		5	Seed bank type IV

(c) Environmental attributes

25	FREQ Frequency in survey	1	<1.0%
		2	1.0-1.9%
		3	2.0-3.9%
		4	4.0-6.9%
		5	7.0% and above
26	STRESS Stress value from ordination	1	Within the first fifth
		2	" " second "
		3	" " third "
		4	" " fourth "
		5	" " top "
27	DIST Disturbance value from ordination	1	Within the first fifth
		2	" " second "
		3	" " third "
		4	" " fourth "
		5	" " top "
28	PH Mode of pH distribution	1	<4.0
		2	4.0-4.9
		3	5.0-5.9
		4	6.0-6.9
		5	7.0 and above
29	MOIST Wetland/total frequency in survey	1	Nil
		2	<0.25
		3	0.25-0.99
		4	1.00-3.99
		5	4.00 and above
30	SHADE Woodland/total frequency in survey	1	Nil
		2	<0.25
		3	0.25-0.99
		4	1.00-3.99
		5	4.00 and above

interrelations between closely similar species than it is that it has followed broader axes of evolutionary progress, the second variant of cluster analysis may be considered more appropriate here. In *optimal* (non-hierarchical) clustering, the analysis sorts the species into a variable number of more or less homogeneous classes. Class properties then become equally important as species properties and the analysis will not necessarily produce the same outcome as would an hierarchical clustering into the same number of classes.

THE UCPE DATABASE

We have sought optimal clusters within a database of **n** = 273 species and **p** = 30 attributes. The species represent the more important components of the herbaceous vegetation of the 2400 km^2 surrounding Sheffield (Grime *et al.*, 1987). The attributes (Table 2) were scored on a uniform (1-2-3-4-5) basis (occasionally 1-3-5) and were selected according to the following four principles: (i) physiological (laboratory-based), biological (taxonomically-based) and ecological (field-based) attributes were each to be included; (ii) attributes of both the vegetative (V) and the regenerative (R) phases of the life cycle were to be included; (iii) no attribute was to be directly derived from any other (to maximise the value of any intercorrelations); (iv) certain environmental (survey-based) attributes of the species, E, were to be included, mainly for reference.

It proved possible to complete the data matrix with no more than 16 per cent of missing data over all. Intercorrelations significant at at least $P < 0.001$ existed both within and between the V, R and E subsets - when expressed as percentages of the possible instances these correlations numbered: within-V, 30; within-R, 22; within-E, 53; V-R, 7; V-E, 30, and R-E, 6.

CLUSTERING POLICY

It is assumed that a globally optimal partition of the **n** species into **g** groups exists, such that species within a group are as homogeneous as possible while species in different groups are as different as possible. Some of the **p** attributes will be more useful than others in defining such a partition, but neither the identity nor the number **k** of the useful attributes is known in advance. Additionally, there is no prior information on the most sensible number **g** of classes in the partition. The purpose of the analysis, therefore, is to find the partition of the species with values of **k** and **g** optimized, the corresponding attributes identified and the group membership of the species determined.

A set of $\mathbf{n}$ species can be partitioned into $\mathbf{g}$ groups in very many ways. A systematic approach to finding the globally optimal partition in the sense defined above is to specify a criterion $\mathbf{V}$ which measures the 'goodness' of any given partition, then to isolate the partition which optimises $\mathbf{V}$. Various criteria and optimizations have been suggested, but the size of the data matrix in the present problem meant that the only computationally viable combination was minimization of the within-group corrected sum of squares of all attributes.

Computation was done using the GENSTAT package. This requires the specification of $\mathbf{g}$, the number of groups into which the species are to be partitioned, and of the $\mathbf{k}$ attributes to be used in the clustering. It then effects the partition using a transfer algorithm (Banfield & Bassill, 1977) and prints the locally optimal value $\hat{\mathbf{V}}_{g,k}$ of the criterion. So much is done automatically, but globally optimal values for $\mathbf{k}$ and $\mathbf{g}$ must be sought by means of a series of repeated analyses. Unfortunately, few guidelines exist in the literature for simultaneous global optimization of $\mathbf{k}$ and $\mathbf{g}$; an appropriate strategy therefore had to be developed.

Marriott (1971) tackled the problem of optimizing $\mathbf{g}$ using the computationally more expensive criterion $\mathbf{V} = |\mathbf{W}|$, where $\mathbf{W}$ is the within-group covariance matrix. Marriott argued that minimization of $\mathbf{g}^2|\mathbf{W}|$ over $\mathbf{g}$ yields an optimal $\mathbf{g}$ because the partition into $\mathbf{g}$ groups of a uniformly distributed population reduces $\mathbf{W}$ by the factor $\mathbf{g}^2$. Data drawn from a population which is strongly grouped, however, produce a disproportionately large reduction in $|\mathbf{W}|$. With the sum of squares criterion, Marriott's arguments suggest that an optimal value of $\mathbf{g}$ involves the function $\mathbf{U} = \mathbf{g}^{\mathbf{a}}\hat{\mathbf{V}}_{g,k}$, with the value of $\mathbf{a}$ determined by the property that $\mathbf{U}$ remains approximately constant over $\mathbf{g}$ when clustering uniformly distributed data. A pronounced minimum of such a function should provide evidence of strong clustering. However, we also require the selection of the $\mathbf{k}$ most useful attributes.

Suppose again that the data are uniformly distributed. Then if $\mathbf{k}$ attributes are used in the clustering, $\hat{\mathbf{V}}_{g,k} \div \mathbf{k}$ will be approximately constant over $\mathbf{k}$ (given an equal contribution of each variate to the total sum of squares). If the data are not uniformly distributed, however, then some attributes will make large contributions to $\hat{\mathbf{V}}_{g,k}$ (the 'poor' attributes), while others will make small ones (the 'good' attributes). The omission of a 'poor' attribute will thus reduce $\mathbf{V}_{g,k} - \mathbf{k}$ by more than the omission of a 'good' one. The contrast may even be more marked for $\hat{V}_{g,k} \div k^b$, where **b** is some constant greater than 1. Hence joint global optimization of **g** and **k** can be approached by the minimisation of $U = g^a\hat{V}_{g,k} \div k^b$ over all possible attributes and group numbers, where **a** and **b** are chosen such that **U** is approximately constant over **g** and **k** when clustering uniformly distributed data.

Suppose that **a** and **b** have been determined. To obtain a globally optimal solution it is necessary to calculate **U** for all possible subsets of the **p** attributes and for all possible numbers **g** of groups, then selecting the **k** and **g** giving the minimum **U**. Since each calculation of **U** involves a very large sequence of cluster analyses, such a scheme is clearly not feasible. We must seek some systematic strategy which reduces the amount of computation while yielding results at least close to the optimum. The following is suggested:

Initially set $k = p$. Let g_{max} be an upper limit to **g**, beyond which a partition into **g** groups is not entertained.

(i) Successively partition the species into **g** groups for $g = g_{max}, g_{max}-1$ ..., 3, 2.

(ii) Examine the curve of $U = g^a\hat{V}_{g,k} \div k^b$ against **g**. Identify the most obvious local minimum U_k (where the curve displays the sharpest 'elbow'). If not clear, this position may be found by comparing the decrements in **U** between **g+1** and **g** with the corresponding decrements between **g** and **g-1**. Denote the value of **g** at this minimum by g_0.

(iii) At $g = g_0$ identify the 'poorest' (least informative) of the **k** attributes by examining the class mean values of each attribute; the 'poorest' has the least variance between the **g** classes.

(iv) Reduce **k** by one, by eliminating the 'poorest' attribute.

(v) Repeat steps (i)-(iv) until the position $k = 2$ has been reached.

(vi) Plot U_k against **k**, and identify the minimal value of **U**. The global optimum is thus at the selected value of **k** and corresponding g_0.

In the above scheme, **a** and **b** must be known in advance. The final step is thus to estimate suitable values for these parameters. It was argued above that these values secure the greatest constancy in **U** when clustering uniformly distributed data. No analytical progress has yet been possible. Accordingly, a Monte Carlo experiment was performed to determine values of **a** and **b** empirically. Nine independent data matrices of size 50 (units) by 10 (attributes) were generated, each entry of each data matrix being an independent uniformly distributed value in the range 1-5 (this being the range of values of the real data used subsequently). The whole scheme outlined above was implemented for each of the nine matrices, giving a total of 729 values of $\hat{V}_{g,k}$ (9 iterations x 9 **ks** x 9 **gs**). If $\mathbf{U} = \mathbf{g}^{\mathbf{a}}\hat{\mathbf{V}}_{g,k} \div \mathbf{k}^{\mathbf{b}}$ is constant for some **a**, **b** for this set of values, then log $\hat{V}_{g,k}$ = const + **blog k** ÷ **alog g**. Hence the values of **a** and **b** which make **U** as nearly constant as possible will be the coefficients of multiple regression of log $\hat{V}_{g,k}$ on log **k** and on minus log **g**. These values turned out to be 0.7 and 1.8 respectively at **p** = 10 attributes, with 96 percent of variance accounted for by the regression. The value of **b** required to give stability to **U** was subsequently found to be sensitive to **p**, drifting from 1.8 to 1.2 as **p** rose from 10 to 30. Values appropriate to the size of the attribute set were selected for each analysis.

CLUSTERING THE UCPE DATABASE

Vegetative, regenerative and environmental attributes have been handled separately and in combination. Here we present analyses of vegetative and reproductive attributes taken separately. In each analysis, two variants of a final optimum classification have been considered, a 'conservative' optimum in which **U** is minimised at $2 \leqslant \mathbf{g} \leqslant 6$ and a 'liberal' optimum in which **U** is minimised at $7 \leqslant \mathbf{g} \leqslant 12$. The class-mean stress and disturbance indices provide a convenient background for a graphical ordination of clusters (Fig. 3), which would otherwise involve making arbitrary selections from within multidimensional structures. Tables 3, 4 and 5 list the notably extreme attributes within each class.

RESULTS FROM THE UCPE DATABASE

Vegetative Attributes

The conservative classification of the vegetative attributes (Table 3) yields three classes, one of which is composed almost exclusively of annual species. The remaining two classes are made up of perennial, mainly polycarpic, species. Both are extremely heterogeneous ecologically and appear to have been separated on the basis of morphological features including leaf dimensions and hairiness. Reference to Figure 3 confirms that the effect of the conservative classification was to separate a compact group occupying the 'ruderal corner' of the triangular model from the remaining perennials. This result reveals no evidence of the predicted distinction between 'competitors' and 'stress-tolerators'.

The liberal classification (Table 4) differentiates seven classes of

Fig. 3. A synopsis of the 'strategic' locations of the 'conservative' and 'liberal' classifications of vegetative and regenerative plant attributes. Points mark the centres of the classes defined in Tables 3, 4 and 5. The triangular base is that constructed by Grime (1984); see the legend to Table 2.

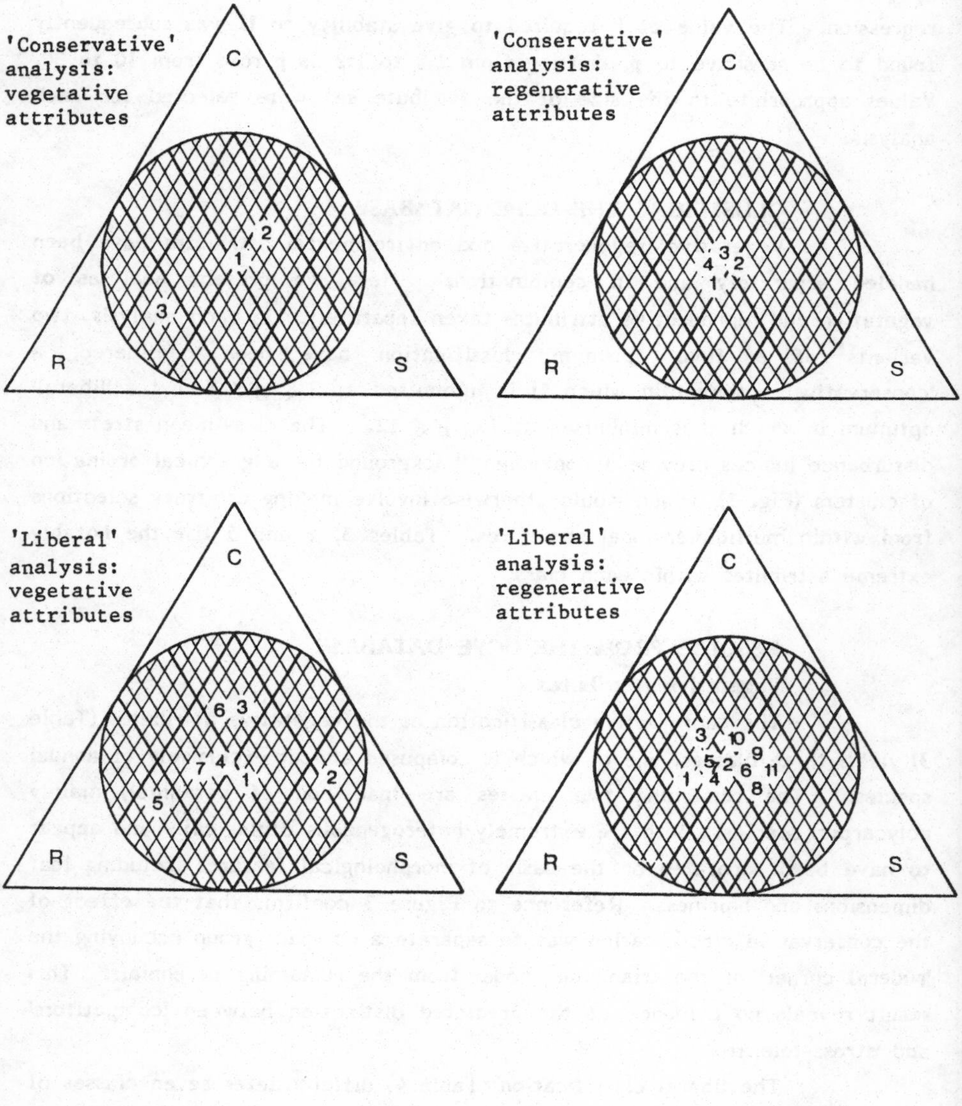

Table 3. 'Conservative' classification based upon vegetative attributes only

Class number	1	2	3
Species in class	94	125	54
Attributes:			
2C DNA content	.	.	.
Canopy height	.	.	.
Lateral spread	.	.	Low
Flowering time	.	.	.
Fl. duration	.	.	.
Life form	.	.	Therophyte
Growth rate	.	.	.
Leaf shape	Broad	.	.
Leaf size	.	.	.
Life history	Polycarpic	Polycarpic	Ephemeral
Leaf phenology	.	.	.
Storage organs	Specialized	.	Seeds
Leaf texture	.	.	Succulent
Leaf hairiness	Very hairy	Not hairy	.
Breeding system	.	.	.
Characteristic species	ACHILLEA MILLEFOL EPILOBIU HIRSUTUM FRAGARIA VESCA HERACLEU SPHONDYL MERCULIA PERENNIS PETASITE HYBRIDUS RANUNCUL REPENS THYMUS PRAECOX URTICA DIOICA VIOLA HIRTA	AGROSTIS CAPILLAR ASPLENIU RUTA-MUR BRACHYPO SYLVATIC CAREX FLACCA EMPETRUM NIGRUM GALIUM VERUM JUNCUS EFFUSUS LOLIUM PERENNE PHALARIS ARUNDINA SEDUM ACRE	AIRA PRAECOX BROMUS STERILIS CHAENOPO ALBUM GALIUM APARINE IMPATIEN GLANDULI JUNCUS BUFONIUS LINUM CATHARTI PAPAVER RHOEAS SENECIO VULGARIS URTICA URENS

Table 5. 'Conservative' classification based upon regenerative attributes only

Class number	1	2	3	4
Species in class	65	64	37	107
Attributes:				
Disp. in space	Limited	Limited	Extensive	Limited
Seed weight	Low	.	Low	.
Initial germin.	High	.	High	Low
Germin. range	.	.	.	.
Germin. dark	.	.	.	.
Germin. mode	.	.	.	.
Seed texture	Smooth	.	.	Smooth
Seed shape	.	Narrow	Narrow	.
Disp. in time	Long	Short	.	.
Characteristic species	AGROSTIS CAPILLAR ANTHOXAN ODORATUM ARABIDOP THALIANA CALLUNA VULGARIS DESCHAMP CESPITOS DIGITALI PURPUREA HYPERICU PERFORAT JUNCUS EFFUSUS MILIUM EFFUSUM RUMEX OBTUSIFO URTICA DIOICA	AIRA PRAECOX ARRHENAT ELATIUS BELLIS PERENNIS BROMUS STERILIS DESMAZER RIGIDA FESTUCA OVINA HOLCUS LANATUS LOLIUM PERENNE POA ANNUA TARAXACU OFFICINA	ATHYRIUM FILIX-FE CHAMERIO ANGUSTIF CREPIS CAPILLAR DRYOPTER FILIX-MA EQUISETU ARVENSE ERIOPHOR ANGUSTIF LEONTODO HISPIDUS SENECIO JACOBEA SONCHUS OLERACEU TYPHA LATIFOLI	ALLIUM URSINUM ANTHRIS SYLVESTR HERACLEU SPHONDYL HYACINTH NON-SCRI LATHYRUS PRATENSI LOTUS CORNICUL RANUNCUL REPENS SANICULA EUROPAEA TRIFOLIU REPENS VIOLA RIVINIAN

Table 4. 'Liberal' classification based upon vegetative attributes only

Class number	1	2	3	4	5	6	7
Species in class	47	28	42	36	52	42	26
Attributes:							
2C DNA content	.	.	.	.	Low	.	.
Canopy height	Low	Low	.	Low	Low	.	.
Lateral spread	.	.	.	.	Low	.	Low
Flowering time	.	.	.	.	.	Late	.
Fl. duration	.	.	.	.	.	.	.
Life form	.	.	.	Therophyte	.	.	
Growth rate	.	.	.	.	.	.	.
Leaf shape	Narrow	Broad	Broad	.	.	Broad	Broad
Leaf size	.	Small	.	.	.	.	Large
Life history	Polycarpic	Polycarpic	Polycarpic	Polycarpic	Ephemeral	Polycarpic	Polycarpic
Leaf phenology	.	Evergreen	.	.	.	Summer only	Summer only
Storage organs	Specialized	.	.	.	Seeds only	.	Specialized
Leaf texture	.	Tough	.	.	Succulent	.	.
Leaf hairiness	Very hairy	Not hairy	Not hairy	Not hairy	.	.	Not hairy
Breeding system	Outbreeding	Outbreeding	Outbreeding	Outbreeding	.	Outbreeding	Outbreeding
Characteristic species	(none)	AVENULA PRATENS CALLUNA VULGARIS CAREX FLACCA DANTHONI DECUMBEN DESCHAMP FLEXUOSA ERIOPHOR VAGINATU FESTUCA OVINA MINUARTI VERNA NARDUS STRICTA VACCINIU VITIS-ID	AGROSTIS STOLONIF ARRHENAT ELATIUS BRACHYPO PINNATUM DRYOPTER FILIX-MA ELYMUS REPENS FESTUCA RUBRA GLYCERIA FLUITANS JUNCUS EFFUSUS PHRAGMIT AUSTRALI PTERIDIU AQUILINU	(none)	AIRA PRAECOX BROMUS STERILIS CHAENOPO ALBUM GALIUM APARINE IMPATIEN GLANDULI JUNCUS BUFONIUS PAPAVER RHOEAS POA ANNUA SENECIO VULGARIS URTICA URENS	ARCTIUM MINUS CHAMERIO ANGUSTIF EPILOBIU HIRSUTUM FILIPEND ULMARIA HERACLEU SPHONDYL MERCULIA PERENNIS PETASITE HYBRIDUS REYNOUTR JAPONICA STACHYS SYLVATIC URTICA DIOICA	(none)

species, four of which (Classes 2, 3, 5 and 6) are ecologically coherent. The group of annuals (Class 1) is retained and, as shown in Figure 3, 'stress-tolerators' emerge in the form of Class 2, low-growing, evergreen species with tough foliage comprising ericaceous shrubs and a variety of grasses and sedges of unproductive habitats. 'Competitors' are represented strongly in Classes 3 and 6, both of which are mainly composed of perennials of large stature and with the potential for extensive clonal expansion. Class 3 contains a high proportion of tall grasses, whilst Class 6 is composed of erect broad-leaved herbs. The remaining classes (1, 4 and 7) are very heterogeneous ecologically and appear to have been formed on the basis of relatively trivial morphological characters.

Regenerative Attributes

Four remarkably coherent classes of species emerge from the conservative classification of regenerative attributes (Table 5). Classes 1 and 3 both consist of small-seeded species but here their similarity ends. Class 3 is made up of species capable of widespread dispersal through the production of numerous buoyant seeds or spores, and clearly corresponds to the regenerative strategy described as W in Table 1. Pteridophytes are prominent in this class which also contains many Compositae and Onagraceae. In marked contrast, Class 1 contains a very high proportion of species in which the small seeds are known to be readily incorporated into a persistent buried seed bank (B_S in Table 1). Although most of the species in Class 1 are those in which the seeds may persist in large numbers in the soil for many years (e.g. Juncus effusus, Rumex obtusifolius) there are also present some species (e.g. Arabidopsis thaliana, Anthoxanthum odoratum) in which the majority of the seeds germinate soon after release and only a small proportion is retained in the seed bank. It is clear that Class 1 corresponds to the seed bank Types III and IV of Thompson & Grime (1979).

Class 2 is an exceedingly homogeneous assemblage of species, mainly grasses, in which the elongated seeds, many bearing prominent awns, lack innate dormancy and exhibit synchronous germination in the autumn (seed bank Type I). A rather more varied set of species appear in Class 4 but, as in Class 2, there is strong evidence of the capacity for regeneration in seasonally predictable vegetation gaps (S in Table 1). In Class 4, however, most of the seeds have a chilling requirement and germinate in the early spring (seed bank Type II). Legumes with hard-coat dormancy are also concentrated in Class 4.

The liberal classification of regenerative attributes produces a large number of classes (11), but many of these appear to relate to relatively minor features of seed morphology.

In both the conservative and liberal classifications (Figure 3) the classes show a strong concentration in the centre of the triangular model which, as stated, relates to attributes of the established phase. This, and the weak intercorrelations of the attributes themselves, provides strong evidence of an uncoupling of strategies of the regenerative phase from those adopted by established plants.

CONCLUSIONS

Despite the limited number of attributes considered and the incomplete nature of the data, the results provide objective support for the existence of recurrent sets of traits (strategies) characteristic of particular ecologies. Evidence consistent with the C-S-R model of primary strategies has been obtained and the analysis strongly suggests that within the local herbaceous flora we have investigated there are four major types of regeneration by seed.

The usefulness of these tests of theories of ecological and evolutionary specialization in plants by multivariate analysis is limited at the present time by the predominance of morphological criteria in our database. Progress in future will depend upon the construction of balanced datasets in which physiological attributes of the kind specified in Table 1 will form an essential component.

We thank all members of the Unit of Comparative Plant Ecology who participated in the field and laboratory studies drawn upon in this chapter. Dr J. G. Hodgson, Mrs. J. M. Fletcher and Mr. A. M. Neal helped particularly in its preparation. All of this work was supported by the Natural Environment Research Council.

REFERENCES

Al-Mufti, M.M. et al. (1977) A quantitative analysis of shoot phenology and dominance in herbaceous vegetation. J. Ecol., 65, 759-91.

Avdulov, N.P. (1931) Karyo-systematische Untersuchungen der Familie Gramineen. Bull. Appl. Bot., 44, Supplement 43, 1-428.

Baker, H.G. (1972) Seed weight in relation to environmental conditions in California. Ecology, 53, 997-1010.

Banfield, C.F. & Bassill, L.C. (1977) A transfer algorithm for non-hierarchical classification. Algorithm AS113. Appl. Statist., 26, 206-10.

Bennett, M.D. (1976) DNA amount, latitude, and crop plant distribution.

Environ. Expl. Bot., 16, 93-108.
Bennett, M.D. & Smith, J.F. (1976) Phil. Trans. R. Soc. Ser. B., 274, 227-74.
Bookman, P.A. & Mack, R.N. (1982) Root interaction between Bromus tectorum and Poa pratensis: A three-dimensional analysis. Ecology, 63, 649-6.
Box, E.O. (1981) Macroclimate and Plant Forms: An Introduction to Predictive Modelling in Phytogeography. Junk, The Hague.
Bradshaw, A. D. (1965) Evolutionary significance of phenotypic plasticity in plants. Adv. Genet., 13, 115-55.
Chapin, F. S. (1980) The mineral nutrition of wild plants. Ann. Rev. Ecol. Syst., 11, 233-60.
Chippindale, H.G. & Milton, W.E.J. (1934) On the viable seeds present in the soil beneath pastures. J. Ecol., 22, 508-31.
Clutton-Brock, T.H. & Harvey, P.H. (1979) Comparison and adaptation. Proc. R. Soc. Lond. B., 205, 547-65.
Coesel, P.F.M. (1982) Structural characteristics and adaptation of desmid communities. J. Ecol., 70, 163-78.
Coley, P.D. (1983) Herbivory and defensive characteristics of tree species in a lowland tropical forest. Ecol. Monogr., 53, 209-232.
Cooke, R.C. & Rayner, A.D.M. (1984) The Ecology of Saprotrophic Fungi: Towards a Predictive Approach. Longman, London.
Crick, J.C. (1985) The Role of Plasticity in Resource Acquisition by Higher Plants. University of Sheffield: PhD Thesis.
Darwin, C. (1859) The Origin of Species by Means of Natural Selection or the Preservation of Favoured Races in the Struggle for Life. Murray, London
Dring, M.J. (1982) The Biology of Marine Plants. Arnold, London.
Furness, S.B. (1979) Ecological Investigations of Growth and Temperature Responses in Bryophytes. University of Sheffield: PhD Thesis.
Gill, D.E. (1978) On selection at high population density. Ecology, 59, 1289-91.
Givnish, T.J. (1982) On the adaptive significance of leaf height in forest herbs. Am. Nat., 120, 353-81.
Gorski, T. (1975) Germination of seeds in the shadow of plants. Physiologia Pl., 34, 342-6.
Gould, S.J. & Lewontin, R.C. (1979) The spandrels of San Marco and the Panglossian paradigm: a critique of the adaptationist programme. Proc. R. Soc. Lond. B., 205, 581-98.
Greenslade, P.J.M. (1972a) Distribution patterns of Priochirus species (Coeleoptera; Staphylindae) in the Solomon Islands. Evolution, Lancaster, Pa., 26, 130-42.
Greenslade, P.J.M. (1972b) Evolution in the staphylinid genus Priochirus (Coeleoptera). Evolution, Lancaster, Pa., 26, 203-20.
Grime, J.P. (1965) Comparative experiments as a key to the ecology of flowering plants. Ecology, 45, 513-15.
Grime, J.P. (1974) Vegetation classification by reference to strategies. Nature, 250, 26-31.
Grime, J.P. (1979) Plant Strategies and Vegetation Processes. Wiley, Chichester.
Grime, J.P. (1983) Prediction of weed and crop response to climate based upon measurements of nuclear DNA content. In: Aspects of Applied Biology, vol. 4, Ed. J. C. Caseley, pp. 87-98. The Association of Applied Biologists National Vegetable Research Station, Wellesbourne, Warwick.
Grime, J.P. (1984) The ecology of species, families and communities of the contemporary British Flora. New Phytol., 98, 15-33.

Grime, J. P. (1985). Towards a functional description of vegetation. In: Population Structure of Vegetation, Eds. J. White & J. Beeftink. Junk, The Hague. (in press).
Grime, J.P. & Hunt, R. (1975) Relative growth-rate: its range and adaptive significance in a local flora. J. Ecol., 63, 393-422.
Grime, J.P. & Mowforth, M.A. (1982) Variation in genome size - an ecological interpretation. Nature, 299, 151-3.
Grime, J.P. et al. (1968) An investigation of leaf palatability using the snail Cepaea nemoralis L. J. Ecol., 56, 405-20.
Grime, J. P. et al. (1981) A comparative study of germination characteristics in a local flora. J. Ecol., 69, 1017-59.
Grime, J.P. et al. (1985) Nuclear DNA contents, shoot phenology and species co-existence in a limestone grassland community. New Phytol., (in prep.).
Grime, J.P. et al. (1987). Comparative Plant Ecology. A Functional Approach to Common British Species and Communities. Allen & Unwin Ltd, London. (in press).
Harper, J.L. (1982) After description. In: The Plant Community as a Working Mechanism. Special Pub. No.1 BES, pp.11-25. Blackwell, London.
Hutchinson, G.E. (1951) Copepodology for the ornithologist. Ecology, 32, 571-7.
Hutchinson, G.E. (1959) Homage to Santa Rosalia or why are there so many kinds of animals? Am. Nat., 93, 145-59.
Jones, R.N. & Brown, L.M. (1976) Chromosome evolution and DNA variation in Crepis. Heredity, 36, 91-104.
Kinzel, W. (1920) Frost und Licht als beeinflussende Krafte der Samenkeimung. E. Ulmer, Stuttgart.
Leps, J.J. et al., (1982) Community stability, complexity and species life-history strategies. Vegetatio, 50, 53-63.
Levin, D.A. & Funderburg, S.W. (1979) Genome size in angiosperms: temperature versus tropical species. Am. Nat., 114, 784-95.
MacArthur, R.H. & Wilson, E.D. (1967) The Theory of Island Biogeography. Princeton University Press, Princeton, NJ.
McNaughton, S.J. et al., (1974) Heavy metal tolerance in Typha latifolia without the evolution of tolerant races. Ecology, 55, 1163-5.
Marriott, F.H.C. (1971) Practical problems in a method of cluster analysis. Biometrics, 27, 501-14.
Noble, I.R. & Slayter, R.O. (1979) The use of vital attributes to predict successional changes in plant communities subject to recurrent disturbances. Vegetatio, 43, 5-21.
Pianka, E.R. (1970) On r- and K-selection. Am. Nat., 104, 592-7.
Pugh, G.J.F. (1980) Strategies in fungal ecology. Trans. Br. mycol. Soc., 75, 1-14.
Ramensky, L.G. (1938) Introduction to the Geobotanical Study of Complex Vegetations. Selkozgiz, Moscow.
Raunkiaer, C. (1934) The Life Forms of Plants and Statistical Plant Geography: being the collected papers of C. Raunkiaer, translated into English by H. G. Carter, A. G. Tansley and Miss Fansball. Clarendon Press, Oxford.
Raven, J.A. (1981) Nutritional strategies of submerged benthic plants: the acquisition of C, N and P by rhizophytes and haptophytes. New Phytol., 88, 1-30.
Reader, P.M. & Southwood, T.R.E. (1981) The relationship between palatability to invertebrates and the successional status of a plant. Oecologia, 51, 271-5.
Roberts, H.A. (1970) Viable weed seeds in cultivated soil. Rep. natn. Veg.

Res. Stn. 1970, 25-38.
Rosen, B.R. (1981) The tropical high diversity enigma - the corals'-eye view. Chance, change and challenge. The evolving biosphere. British Museum and Cambridge University Press.
Salisbury, E.J. (1942) The Reproductive Capacity of Plants. Bell, London.
Shepherd, S.A. (1981) Ecological strategies in a deep water red algal community. Botanica mar., 24, 457-63.
Sibly, R.M. & Grime, J.P. (1986) Strategies of resource capture by plants - evidence for adversity selection. J. Theor. Biol., 118, 247-50.
Stebbins, G.L. (1956) Cytogenetics and evolution of the grass family. Am. J. Bot., 43, 890-905.
Taylorson, R.B. & Borthwick, H.A. (1969) Light filtration by foliar canopies; significance for light-controlled weed seed germination. Weed Sci., 17, 48-51.
Thompson, K. & Grime, J. P. (1979). Seasonal variation in the seed banks of herbaceous species in ten contrasting habitats. J. Ecol., 67, 893-922.
van der Valk, A.G. & Davis, C.B. (1976) The seed banks of prairie glacial marshes. Can. J. Bot., 54, 1832-8.
Wilbur, H.M., et al. (1974) Environmental certainty, trophic level and resource availability in life history evolution. Am. Nat., 108, 805-17.
Woolhouse, H.W. (1981) Aspects of the carbon and energy requirements of photosynthesis considered in relation to environmental constraints. In: Physiological Ecology: an Evolutionary Approach to Resource Use, Eds. C. R. Townsend & P. Calow. Blackwell, London.

ADAPTIVE CHARACTERISTICS OF LEAVES WITH SPECIAL REFERENCE TO VIOLETS

O.T. Solbrig

INTRODUCTION

Vegetative and reproductive characteristics of plants are presumably adaptive in the sense that they allow survival and reproduction of the individual in some, but not all environments (Mason & Langenheim, 1957; Solbrig, 1981a). If so, species in different environments should possess different characters. Habitat partitioning among closely related species possessing different characteristics has been reported along gradients of environment (Marshall & Jain, 1968, 1969; Abrahamson & Gadgil, 1973; Pickett and Bazzaz, 1976, 1978; Werner & Platt, 1976; Hickman, 1977; Werner, 1979; Grace & Wetzel, 1981), temperature (Thomas & Dale, 1976), light (Harper & Clathworthy, 1963), disturbance (Solbrig & Simpson, 1974, 1977), and soil properties (Parrish & Bazzaz, 1976, 1979; Sharitz & McCormick, 1973).

Though green land plants harvest essentially the same resources (light, water, and a variety of nutrients), these resources are not of uniform quality nor are they harvested in the same way by all species. There are well known physiological and morphological trade-offs that can result in the coexistence of species even though they are harvesting the same or similar resource. So, for example, in tropical savannas, trees and grasses coexist. The former have deep roots that extract water and nutrients from deep soil layers; the latter have most of their roots in the upper 20-60 cm of the soil (Sarmiento, 1984; Sarmiento et al., 1985).

Leaves are the manufacturing centres, the factories of plants. In them carbon dioxide and water get converted into carbohydrates using light energy. Here too, some of the machine-tools of the leaf are built, the enzymes, the membranes, the organelles. Leaf sizes and shapes can be used as climatic indicators (Raunkier, 1934). Leaf form is correlated with environmental characteristics, both climatic and edaphic, as well as with other plant characters, especially root characteristics. A number of these changes have been studied in detail and models have been proposed that predict the optimal leaf morphology for different environmental situations (Gates & Papian, 1971; Parkhurst & Loucks, 1972; Taylor, 1975; Givnish & Vermeij, 1976; Ehleringer, 1976; Ehleringer et al., 1976; Orians & Solbrig, 1977; Givnish, 1979).

This paper deals with the characteristics of the factory and not with the processes of manufacturing. So as not to carry the analogy too far, what I am asking specifically is, what are the design characteristics of leaves that affect photosynthate production? The problem reduces to finding those leaf characteristics that (1) can maintain leaf temperature close to the optimal temperature of photosynthesis, during the longest possible span of time in a day, and repeat such for as long as the leaf is functional; (2) keep leaf water potentials from becoming too negative; and (3) keep the chloroplasts supplied with CO_2. The problem is complex because (a) the optimal temperature for photosynthesis can be adjusted by the individual plant within certain limits; (b) reduced leaf water potential affects photosynthetic performance differently in different species; and (c) species vary in their requirements for CO_2. Consequently it is very difficult to set up general hypotheses regarding optimal leaf shape. Instead we possess general rules, based on biophysical, biochemical, and mechanical principles. These questions have been addressed among others by Gates et al., 1968; Gates & Papian 1971; Parkhurst & Loucks 1972; Taylor 1975; Mooney & Gulmon 1979; Givnish 1979; Solbrig 1981a, Givnish 1984 and Bloom et al., 1985. Because the subject has been dealt with in detail before, only the barest outlines are presented here. The interested reader is referred to the literature cited above.

First I will present a list of some of the morphological and physiological characters that are likely to vary in a leaf (the dependent variables) and of the principal environmental factors that affect leaf function (the independent variables). I then present an overview of some of the reasoning used in constructing general models that predict optimal values of the dependent variable as a function of the independent variable. Some of the predictions are then illustrated with examples from work by myself and coworkers with Viola.

THE PRINCIPAL LEAF CHARACTERISTICS: THE DEPENDENT VARIABLES

The first and most obviously varying characteristic of leaves are their physical dimensions; length, width, and thickness. Leaf size varies both between taxa and within taxa, and within individuals. Factors that affect leaf dimension are water, light, and nutrients.

A second varying characteristic of leaves is shape. Shape is more illusive to evaluate quantitatively (see however Kincaid & Schneider 1983), and usually is characterized qualitatively (DeCandolle, 1880; Anonymous, 1962). Shape can be broken up into subsidiary characters. Of these the most important are whether the leaf is entire or dissected (lobed, pinnatifid. etc.) (see DeCandolle, 1880), the type of margin that it possesses (entire, toothed, lobed, etc.), and the insertion of the petiole.

Another important morphological character is the type of leaf surface, whether smooth, or rough, glabrous or hairy, shiny or dull. These characters can affect the energy the leaf absorbs from the sun (shiny or hairy leaves reflect more light than dull or glabrous leaves); and the amount of water vapour that the leaf exchanges with the atmosphere because of the effect of surface on the leaf boundary layer.

Among anatomical characters the most important is the shape, size, position, and density of stomata, which are the most important regulatory mechanism of the leaf. By controlling gas exchange, they indirectly affect leaf temperature, hydrature, and even nutrient uptake (which is related to the transpirational stream). Other anatomical characteristics of importance are the type, size, shape, position, and density of conducting bundles, which affect leaf irrigation, and mechanical characteristics. Also the type of photosynthesis (whether C3 or C4); and the number, size, and relative position of palisade and spongy parenchyma, which affect light transmission, and gas exchange within the leaf. The relative percentage of fibres, and hard tissues in relation to total weight of the leaf affect palatability of the leaf to herbivores and determines the quality of litter.

Leaves also possess waxy cuticles, which can be quite massive in some instances. The physical and chemical characteristics of these cuticles affect gas exchange through the surface of the leaf, and can affect palatability of the leaf to herbivores. Very thick cuticles can occasionally also interfere with stomatal function. The cells of the leaf, either all or only some specialized ones such as latex tubes or epidermal cells, may be filled with toxic or unpalatable chemicals to deter herbivores. Such chemicals may interfere directly or indirectly with photosynthate production. They will also affect decomposition, and indirectly soil flora and fauna, and soil nutrient status (Hale et al., 1978).

Another leaf variable is longevity. Leaves are usually ephemeral structures, that are renewed at periodic intervals. However at least in one case, that of Welwitchia mirabilis, the leaves, here cotyledonary, are as long lived as the plant itself. Other plants renew their leaves only once a year, such as Curatella americana from the South American savannas.

THE PRINCIPAL ENVIRONMENTAL FACTORS : THE INDEPENDENT VARIABLES

Radiation (both short, visible, and long wave radiation), air temperature, air and soil humidity, and wind are the principal environmental factors that affect performance of leaves with given characteristics. To this list must be added herbivory as an additional important selective factor.

Plants can utilize only wavelengths in the 400-700 nm range. Since plants do not possess a filtering mechanism that allows them to absorb only in the desired range, excess energy must be dissipated. Air temperature affects leaves by affecting the radiative environment. Air humidity affects the rate of transpiration, i.e. the loss of water from the leaf. If air humidity is very high, the flux of water vapour from the leaf to the atmosphere may be too slow to dissipate the absorbed energy through evaporative cooling; if air humidity is very low, the loss of water vapour from the leaf may be higher than the capacity of the plant to replenish it from the soil water solution. Since air and soil water humidity are usually correlated, evaporative demand will tend to be highest in environments with dry soils, and vice-versa. Finally, wind speed is an important independent environmental variable affecting leaf form and function. Increasing wind speed can increase transpiration by decreasing the boundary-layer resistance (although the relation is not linear), and also increase loss of water from the soil by increasing turbulence and mixing of the air column above the soil, and thereby decreasing boundary layer resistance above the soil and vegetation (Bruning, 1970). Increased wind, by cooling the leaf can also work in the opposite direction, leading to a decrease in transpiration.

A final independently varying environmental factor affecting the characteristics of leaves is herbivory. Plants use many types of defenses against potential herbivores: short lived leaves, hard to digest leaves due to high fibre or tanin contents, and presence of toxic chemical compounds. The kind of defence depends on a variety of factors which will not be considered here (Mooney & Gulmon, 1982; Mooney et al., 1983).

In turn, chemical and physical properties of leaves have an effect on the type of decomposer organisms that attack the litter, the speed of decomposition, and the manner in which nutrients are made available to plants.

LEAF SIZE AND FUNCTION

A leaf can be compared with a solar panel. Basically it is a surface that absorbs light which is converted by the chloroplasts into carbohydrates. Since only 1 to 3% of the incident light energy is used in photosynthesis, the leaf must dissipate most of the remainder. Some of it is reflected (especially when the leaf is shiny or hairy), some of it is transmitted (particularly in thin leaves), and the remainder is absorbed. In the majority of leaves more than 50% of the incident light is absorbed (Birkebak & Birkebak, 1964), and must be dissipated by evaporative or convective cooling, or re-radiated. The relation between evaporation (transpiration), convection and radiation is expressed by the leaf energy balance equation as follows:

$$R_n - H - LE \pm J = 0$$

where R_n = absorbed radiation (in $Wm^{-2} K^{-1}$); H = sensible heat; L = latent heat of vaporization (2450 Jg^{-1}); E = transpiration rate; and J = rate of change in stored heat flux density.

In most environments, the primary loss of heat is through evaporation and convection, the amount lost by re-radiation is small in comparison as leaves normally stay close to air temperature. Evaporation consumes large quantities of water from the leaf that must be replaced by roots from the soil solution.

Photosynthetic rates increase with temperature like any other chemical reaction. Because the increase in leaf temperature with increased size is steeper than the increase in the boundary layer resistance, photosynthesis and transpiration increase at different rates with increase leaf size. As leaf temperature increases, the photosynthetic rate will eventually level off when the rate of CO_2 diffusion reaches its maximum value. Beyond that point, as leaf temperature increases, transpiration and respiration increase without any photosynthetic gain. Beyond a certain point photosynthetic rates actually go down. Higher transpiration rates demand a greater supply of water to the leaf. Otherwise, leaf water potential drops, and this has a depressing effect on photosynthesis (Fig. 1). To maintain this greater supply of water, photosynthate has to be devoted to the production of more roots and conducting tissue, thereby reducing the photosynthetic energy and materials that are available for reproduction, protection or further leaf production (Solbrig, 1981a).

Leaf size and shape determine the boundary layer of the leaf in still air. In turn, heat loss by conduction and convection as well as transpiration are affected by the magnitude of the boundary layer. This effect was first investigated by Raschke (1960) and Gates (1965). The larger the boundary layer (for a given environment), the greater the resistance, and therefore the equilibrium temperature of the leaf, for a given set of environmental circumstances. The boundary layer is affected also by the wind speed. The approximate thickness of the boundary layer next to the leaf is (Nobel, 1983)

$$D = a \cdot l/v$$

where D = thickness of the boundary layer (in mm); l = mean length of the leaf in the down wind direction (in m); v = ambient wind speed (in ms^{-1}); and a = factor for the effective boundary layer thickness adjacent to a flat plate

Fig. 1. Benefit and cost curves of photosynthesis (full line) and transpiration (dotted line) for leaves in a sunny environment as a function of effective leaf size, (Givnish, (1979). The different curves indicate more sunny (1) or less warm (2). The optimal size for a given environment is where benefit most exceeds costs (Givnish, 1979).

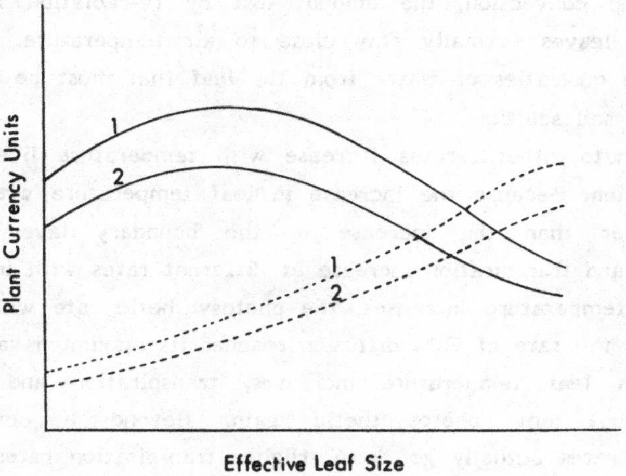

(Schlichting, 1968; Monteith, 1981). This factor, which for a flat plate is 6, has been shown in wind tunnel experiments to be closer to 3.7 for leaves (Gates, 1980). The exact value is affected by leaf shape, surface characteristics, and leaf flutter. It must be obtained empirically.

Rate of transpiration depends on both environmental characteristics and leaf properties. It can be represented in its simplest formulation as a diffusion process as follows :

$$Tr = H^1 - H^a / r_s + r_b$$

where H^1 = air humidity in the inside of the leaf (water vapour density in gm^{-3}; H^a = water vapour density of the air; r_s = stomatal resistance (sec. cm^{-1}) and r_b = boundary layer resistance.

It was already mentioned that the boundary layer is a function of the size, shape, and surface of the leaf. Likewise, the density, size and characteristics of stomata have an effect on transpiration and leaf temperature. Stomata open and close during the day, thereby changing the value of the transpiration rate and affecting leaf temperature. Ideally, stomata should vary with environmental conditions in the manner of a variable resistance (opening more when absorbed radiation increases, closing slightly with increasing wind speeds that lower the boundary layer resistance) to maintain the leaf close to the optimal temperature for photosynthesis. However, stomata also control the rate of CO_2 diffusion into the leaf, so that by closing they may affect photosynthesis more negatively than they would by remaining open and permitting a less than ideal temperature for photosynthesis.

Another factor that needs to be considered is leaf hydrature, since a drop in the leaf water potential also has a negative effect on photosynthesis (Boyer, 1971). Consequently, in some circumstances, closing stomata to maintain leaf hydrature may result in more photosynthate production than leaving them open, even allowing for the corresponding drop in the rate of CO_2 flux, and/or a change in leaf temperature from the photosynthetic optimum.

Since small leaves stay closer to air temperature than large leaves even when stomata are closed, these small-leaved plants are expected to be the norm in environments where stomata are closed for part of the day due to water stress from drought or high evaporative demand (Parkhurst & Loucks, 1972; Givnish, 1979, 1984).

Since leaves are like solar panels, small leaves intercept less light and consequently produce less photosynthate than larger leaves.

Consequently small leaves may be less effective than large leaves. However, in comparing a leaf to a solar panel one must be careful not to commit the common fallacy of comparing single large leaves with single small leaves. Rather one must ask whether a single leaf is analogous to a collection of small leaves of comparable area. The question also must be framed in terms of net gain, taking both costs and benefits into account, and not just in terms of gross income.

The gain for the plant is clearly photosynthate production. The costs are basically three: (1) costs of construction; (2) costs of maintenance (respiration); and (3) the costs of transpiration.

High photosynthetic rates will be sustained only when light flux density is saturating the photoreceptors, temperature is close to optimal, leaf water potential is close to zero and the rate of carbon dioxide diffusion is high. Light and carbon dioxide fluxes, and leaf temperature and water potential are mutually interdependent and their interactions are complex. Therefore there is probably more than one set of leaf characteristics that provides adequate supplies of light, water and carbon dioxide to photosynthesizing leaves. To find out what those alternatives are, and to learn which of them may be favoured by natural selection in a particular environment, the relevant state space, constraints and cost functionals of the system must be identified and described (Solbrig, 1981a).

Knowledge of the great many variables that affect photosynthate production and their interactions is not yet precise enough to predict the optimal characters of leaves <u>exactly</u> for a given species in a given environment. It is not even known whether there is a single optimal leaf, or several optimal solutions. Furthermore, since the relevant environmental characteristics (light, temperature, humidity, and wind) vary in time and space, and some of the significant leaf parameters (stomatal resistance, leaf water potential) also vary in time and space, while others (leaf size, thickness, leaf angle, and hairiness) are developmentally plastic, and therefore vary somewhat from leaf to leaf and from plant to plant within a species, it is unlikely that an optimal leaf (in the classical sense) exists, or has any physiological meaning. But since leaf characteristics affect photosynthate performance in different environments, they are undoubtedly a factor in determining the distributional limits of a species, and its fitness. In this sense, the study of leaf form and function can be very important in an evolutionary and ecological sense.

EVOLUTIONARY AND ECOPHYSIOLOGICAL STUDIES ON VIOLETS

New England violets, provide an excellent opportunity to test some of the hypotheses that have been proposed on morphological and physiological trade-offs between species of high light (open fields) and low light (forest floor) environments. Here we report convergence in morphological and physiological characters of two pairs of closely related species of high and low light environments: Viola fimbriatula, and V. sororia; and V. lanceolata, and V. blanda.

The Species

Two species, V. blanda and V. lanceolata belong to subsection blandae of section Plagiostigma, the so called "stemless white" violets. They are characterised by their white chasmogamous flowers and slender stolons. They form loose patches containing a few tens to several hundred plants, and tend to be locally dominant where they grow. Viola lanceolata has n=12, which is the base chromosome number of the section, while V. blanda is an aneuploid with n=22. Hybrids, both natural and artificial, have been reported between these species. The other two species, V. sororia and V. fimbriatula belong to subsection novae-angliae of section Plagiostigma. They too are stemless violets, characterised by their blue flowers, and short, stubby rhizomes. They reproduce asexually and much less frequently than the stemless white violets, and tend to occur also in patches, but of lower density than those formed by white flowered species. They are found preferentially in disturbed sites. Both these species have 27 pairs of chromosomes and natural, fully fertile, hybrids are not infrequent.

Both V. blanda and V. sororia are found in New England growing in the forest floor, the former preferentially in moist places, often in areas with standing water in spring, such as depressions or drainage channels. Viola sororia is found mostly in drier sites, especially following minor soil disturbance. Both of these forest species possess cordate leaves up to 50mm long by 50 mm wide by 30 μm thick, held horizontally on erect petioles. Early spring leaves are smaller and are presented at a lower height than mid-summer leaves. They possess stomata only on the lower surface of the leaf. Viola lanceolata and V. fimbriatula grow preferentially in open places such as meadows and forest clearings. The former prefers moist places with standing water in spring, the latter grows in a variety of soil moisture conditions including very dry sandy soils with little vegetation. Viola lanceolata possesses narrow, lanceolate leaves, up to 10 cm long, by 2.5 cm wide, and 35 μm thick. The leaves are held close to vertical on short, thick petioles, 2-3 cm long. Viola fimbriatula also has elongate leaves, 10-15 cm long, by 2-4 cm wide, and 50 μm thick, held at varying angles, from almost horizontal to vertical. Both species are amphistomatous.

Related species are similar in reproductive characteristics and in their preference for microsites within their respective environments: the white flowered species grow in slightly wetter, the blue-flowered in drier microsites; on the other hand the two forest species and the two meadow species appear to converge in their foliar characters.

Demographic studies (Solbrig et al., 1980; Newell et al., 1983; Solbrig et al., in prep.) reveal that all four species have similar general phenologies and life histories. All are short-lived perennial species. The half life of ramets of the stoloniferous species (5-7 years) is shorter than that of the rhizomatous ones (10-15 years), but the half life of genotypes of the former could be longer on account of their greater extent of vegetative reproduction. All four species produce chasmogamous flowers in spring (April-May), followed by cleistogamous flowers and fruits in summer. Plants go dormant in September or early October. Survivorship of seedlings is a function of size attained (i.e. of growth rate) (Cook, 1979, 1980; Solbrig et al., in prep.). Experimental studies with both V. sororia and V. fimbriatula (Solbrig, 1981b; Solbrig et al., in prep.) indicate that additive genetic variance for growth rate in seedlings is very low, and that most of the variability between seedling growth rate and seedling survivorship is attributable mostly to the environment. Survivorship and fecundity of adults is equally size dependent. The 10% largest plants in the population produce between 50% and 80% of all seeds produced in a year in a population, which amounts to a rather strong regime of selection. Experimental studies with V. sororia (Antlfinger et al., 1985) indicate that there is low heritability for growth rate, but significant heritability for various leaf and reproductive characters. From the demographic analysis it is clear that any variation that enhances growth, especially among young plants, is translated into a greater fitness advantage. We now ask, in what ways do the foliar differences between field and forest species enhance photosynthesis and growth.

Outline of Working Hypothesis

The hypothesis that was tested is that the evolution of these four species of violets has involved the occupation of different habitats and that this has been accompanied by morphological and physiological changes (especially in leaf characters) that have better adapted the species to these habitats. The null hypothesis is that the morphological changes that are observed are not necessarily adaptive, but are simply a byproduct of isolation. The test of the hypothesis consists in showing that populations grow better (i.e. less mortality, higher fecundity) in their native habitats than in foreign habitats and in presenting ecological and physiological reasons for this behaviour.

More specifically, I contend that two environmental factors have been paramount in the evolution of these species: soil-water availability and light availability. The blue-flowered species are more drought adapted than the white-flowered species, while the field species of each pair are in turn more adapted to a high-light environment than the forest species. This is reflected in the morphology of the plants: the R/S ratios are higher in the blue-flowered species compared to the corresponding white-flowered species in their respective environments (forest or field) and the field species compared to the forest species within their own groups (blue- or white-flowered). Field species differ from forest species also in a number of leaf characters. The significance of these leaf character differences is what I wish to emphasise in this paper.

The study consisted of three steps, which will be discussed briefly (detailed presentation of observation and experimental results will be published elsewhere) :

1. Study of the array of microclimates occupied by these four species, and of their physiological behaviour (transpiration, photosynthesis).
2. Evaluation of the ecological and adaptive significance of the morphology and physiological behaviour of leaves under different environments.
3. Test of the predictions developed in 2 through observation and transplant studies.

Microclimate and Physiology

Microclimatic measures were taken as in standard techniques to be discussed elsewhere (Solbrig et al., in prep.). Stomatal resistance was measured in the field with a Licor LI-60 diffusive resistance meter or with a LI-1600 steady-state psychrometer according to the size of the leaves. Photosynthetic curves were obtained in the laboratory using an open infra-red gas analyser system (Mooney et al., 1971). Finally, leaf nitrogen values were obtained using a Microkjeldahl apparatus.

Morphological information obtained in the field and through the season on ten marked plants was : (1) number of leaves; (2) leaf blade dimensions (length, width, thickness); (3) petiole length, and (4) angle of the blade with the soil plane.

The microclimatic information was used in constructing models of leaf energy budgets, transpiration and photosynthesis that will be presented elsewhere (Solbrig et al., in prep.). Here we report on the major environmental and leaf characteristics separating the study sites and the species under investigation.

The major differences in the sites is in relation to light, water, and to a lesser degree temperature. The forest sites (particularly V. blanda) have a much lower level of illumination than the open field sites, which received an unobstructed radiational stream. Maximum instantaneous values for unshaded sensors in the open field were 150-180 nE cm^{-2} s^{-1}. In the shade, at the beginning of the growing season, only 30% of the light stream penetrated to the forest floor; as the tree canopy closed, only 10% or less reached the forest floor (approximately 10 nE cm^{-2} s^{-1}. However, this low light level was punctuated by sun flecks of short duration (10 minutes or less) (Curtis & Kincaid, 1984).

All sites had higher levels of soil humidity at the beginning of the season as a result of snowmelt and winter storage, and dried up during the season. The blanda site was considerably wetter than the other three sites, and the lanceolata site was wetter than the fimbriatula site. Air temperatures were similar in comparable days at the two forest sites and the two field sites, but the forest sites were up to 5°C cooler than the field sites, depending on environmental conditions.

Viola fimbriatula had the largest average leaf surface/plant followed by V. blanda, V. lanceolata, and V. sororia. V. fimbriatula also had the greatest average leaf weight/plant, followed by V. lanceolata, V. blanda and V.sororia in that order (Figs. 2 & 3). That is, V. lanceolata produced more leaf mass per unit surface than V. blanda. V. fimbriatula also had the thickest leaves, followed by V. lanceolata, V. blanda, and V. sororia (Fig. 4). V. blanda produced its leaves on the largest petioles, followed by V. sororia, V. lanceolata and V. fimbriatula (Fig. 5). The average leaf angle formed by the leaves with the soil plane was close to 0° in populations of the forest species but was above 40° in V. lanceolata and varied from 20° at the

Fig. 2. Average leaf surface in cm2 (A) and average leaf weight of leaves in gms. (B) of individual plants of V. fimbriatula (solid line) and V. sororia (dotted line) at the Concord Field Station during 1983.

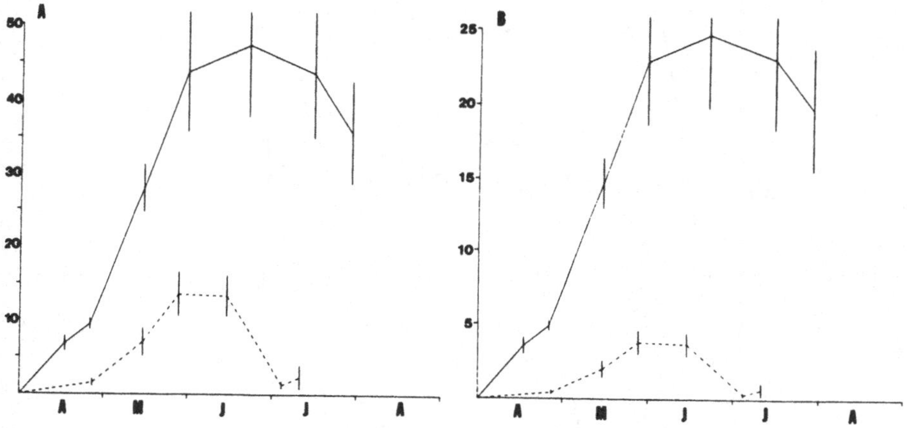

Fig. 3. Average leaf surface in cm2 (A) and average leaf weight of leaves in gms. (B) of individual plants of V. lanceolata (solid line) and V. blanda (dotted line) at the Concord Field Station during 1983.

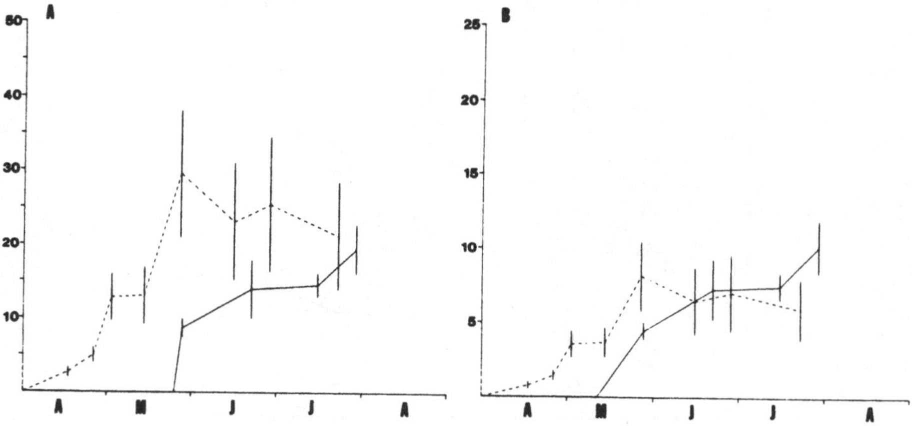

Fig. 4. Average leaf thickness in hundreds of mm. from April to August in a population each of V. fimbriatula (F), V. lanceolata (L), V. blanda (B), and V. sororia (S) at the Concord Field Station in 1983.

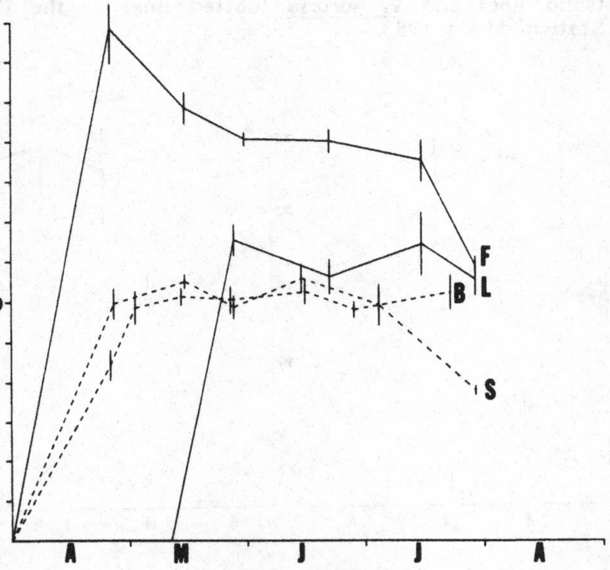

Fig. 5. Average petiole length (in cm) from April to August in a population each of V. blanda (B), V. lanceolata (L), V. sororia (S), and V. fimbriatula (F) at the Concord Field Station in 1983.

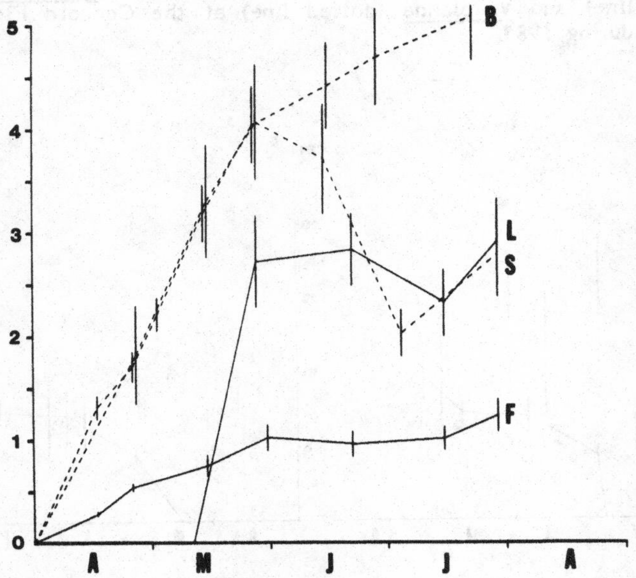

beginning of the season to 40° at the end of the season. Individual leaves varied from close to horizontal to almost in V. fimbriatula vertical.

Evaluation of the Ecological and Adaptive significance of the Morphology and Physiological Behaviour of Leaves.

The study confirmed earlier results that indicate great similarity in the photosynthetic behaviour of these species (Curtis, 1979, 1984). All four species have low photosynthetic rates, low compensation and saturation points, and high quantum yields. Significant differences in photosynthate production between field and forest species were however uncovered and these can be linked directly to the morphological differences that exist between the species.

The photosynthetic rate of field species expressed on a surface (but not on a weight basis, Fig. 6) is higher than that of the forest species (Curtis, 1984; Solbrig et al., in prep.). This difference is due to the thicker leaves of the field species. In a high light environment, more layers of cells can become light saturated than in a low light environment, especially in species with low light compensation points. Thick leaves also lose less water per unit weight than thin leaves, since transpiration is a function of leaf surface but not weight. Consequently thick leaves will be favoured in high light environments.

The effect of differences in leaf shape, size, and stomatal conductance, on leaf temperature, and indirectly on transpiration and photosynthesis, was estimated using a computer model that calculates photosynthesis, transpiration and leaf temperature given air temperature, wind speed, air humidity, radiation, stomatal resistance, and leaf dimensions. Calculated values were then compared with actual measured values to verify the model. The model calculates leaf temperature very accurately, with less than one degree error. Using this model it was established that leaves of V. blanda that are normally found in a low light environment would warm up to 10°C above air temperature if confronted with a high light environment such as that encountered by the field species V. fimbriatula and V. lanceolata, principally on account of their high stomatal resistance. On a hot day in New England when air temperatures can reach up to 35°C in an open field, it can mean leaf temperatures well over 40° C, which could be close to lethal temperatures. This may explain the absence of this species from high radiation environments. When plants of V. blanda are grown in growth chambers under higher light conditions (1000 $mE.cm^{-2}.sec^{-1}$) than in nature

Fig. 6. Photosynthetic rates (expressed in $nMols.cm^{-2}.sec^{-1}$ (circles), or $nMol.mg^{-1}.sec^{-1}$ (squares) for all four species from April to August. Note that while the rates are different between species and dates when expressed on a per surface basis they are very similar when expressed on a per weight basis. The difference is due to differences in leaf thickness.

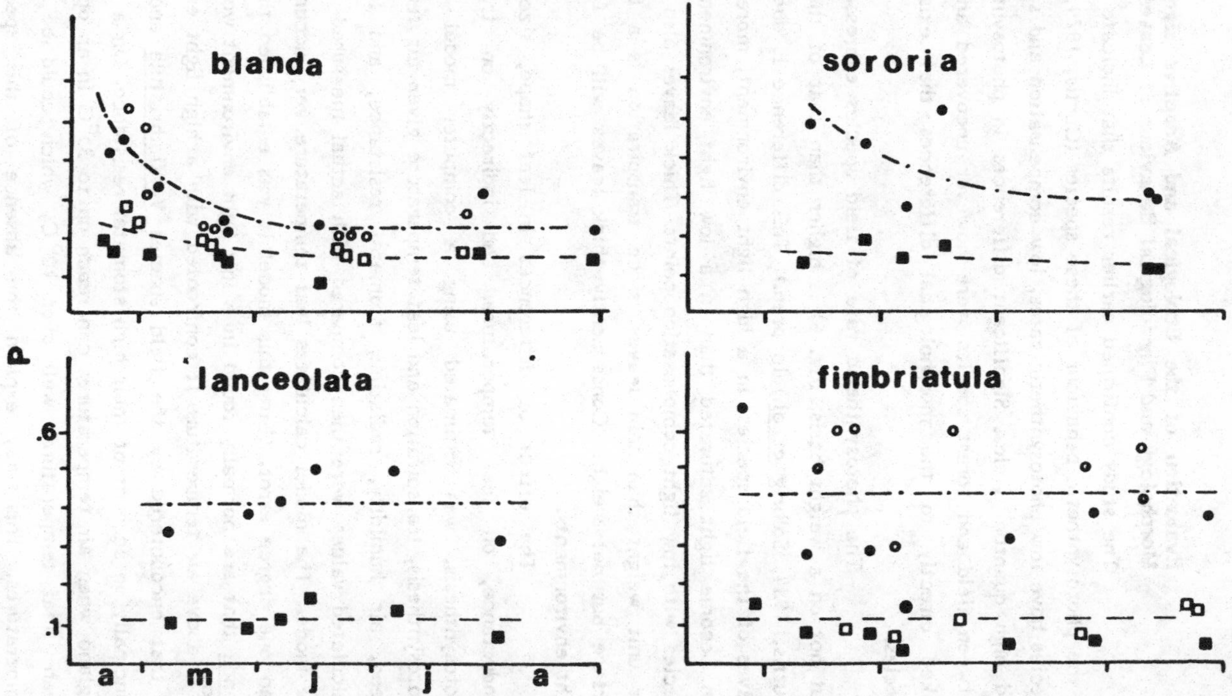

they photobleach and die within a period of 2-3 weeks (Curtis, 1979, 1984). The normal life span of a leaf of V. blanda in the field is 3-4 months. Photobleaching is probably brought about by the high leaf temperatures.

Independent confirmation of the inability of leaves of V. blanda to maintain leaf temperature close to air temperature in high radiation environments, is provided by measurements of leaf temperatures of leaves subjected to long duration sun flecks in the forest (five or more minutes). Leaf temperatures that can be up to 5°C below air temperature when in the shade, quickly increase their temperature to values above air temperature. In one instance a temperature of 39.8°C was recorded.

Using this same model, the effects of changes in leaf angle and azimuth can be calculated. For a leaf exposed to a high light environment such as V. fimbriatula, changing the leaf angle and azimuth has a greater effect on transpiration than on photosynthesis, and therefore water use efficiency increases with increasing angle for most azimuth angles (Fig. 7). In a low light environment, the opposite is true, especially in a forest environment, where early morning and late afternoon radiation is very low on account of light interception by tree trunks. In such environments, changes in leaf angle have an immediate negative effect on photosynthesis but a minimal effect on transpiration. Consequently, water use efficiency decreases if the leaf deviates significantly from horizontal.

Test of the Predictions

The analysis just described makes the prediction that each species should perform best in its native environment. The prediction was tested at the Harvard Forest with one population each of the following three species : V. sororia, V. fimbriatula and V. blanda. These three populations grew within 300 m of each other: V. sororia in a dry part of the woods, V. blanda in a wet depression fifty meters away, and V. fimbriatula at the edge of the woods, in a small clearing by a road, about 300 m away from the first two. Fifty plants of each species were planted in a 10 x 15 grid within each of the three sites on June 22, 1982, and their fate followed during a period of three years.

Table 1 and Fig. 8 summarise the results. The plot that contained originally only plants of V. fimbriatula had the highest survival rate (54 out of 150 plants), the sororia plot had the second highest survival rate (30 out of 150) and the blanda plot the lowest (17 out of 150). After three years Viola blanda was restricted to its original site (excluding one plant growing in

Fig. 7. Photosynthate production (in $mg.day^{-1}$) (squares), transpiration (in $gm.day^{-1}$) (circles), and water use efficiency (P/Tr) (triangles) estimated for leaves of V. fimbriatula with different angles of elevation over the horizontal (considered as 0°) and azimuth.

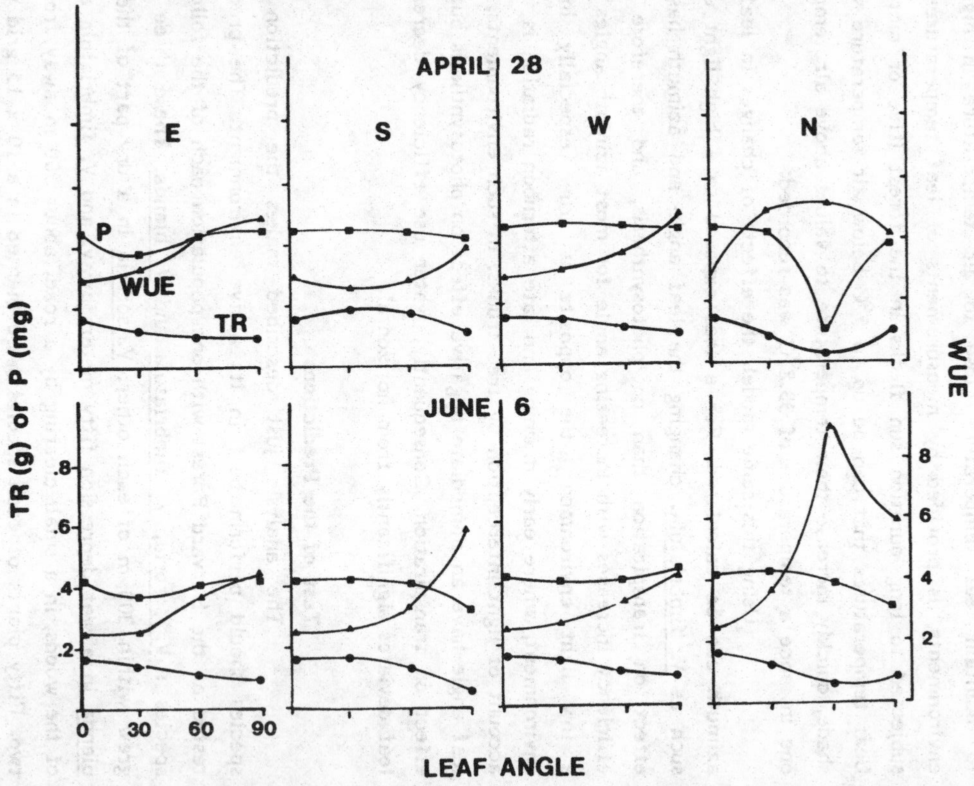

Table 1. Survivorship of transplants

site	fimbriatula	sororia	blanda	Total
species				
fimbriatula	31 (57.4%)	13 (43.3%)	0 (0%)	44
sororia	23 (42.6%)	16 (53.3%)	3 (17.6%)	42
blanda	0 (0%)	1 (3.3%)	14 (82.4%)	15
	54 (100%)	30 (100%)	17 (100%)	101

Chi Sqr. for 4 d.f. = 75.989 ***   Coef. of Conf. = 0.655

Fig. 8. Per cent survival of 150 plants (50 at each site) of V. fimbriatula (F), V. sororia (S), and V. blanda (B), transplanted from and to three sites at the Harvard Forest in Petersham, Mass., where originally only grew V. fimbriatula (Fimb), V. sororia (Sor), and V. blanda (Blan).

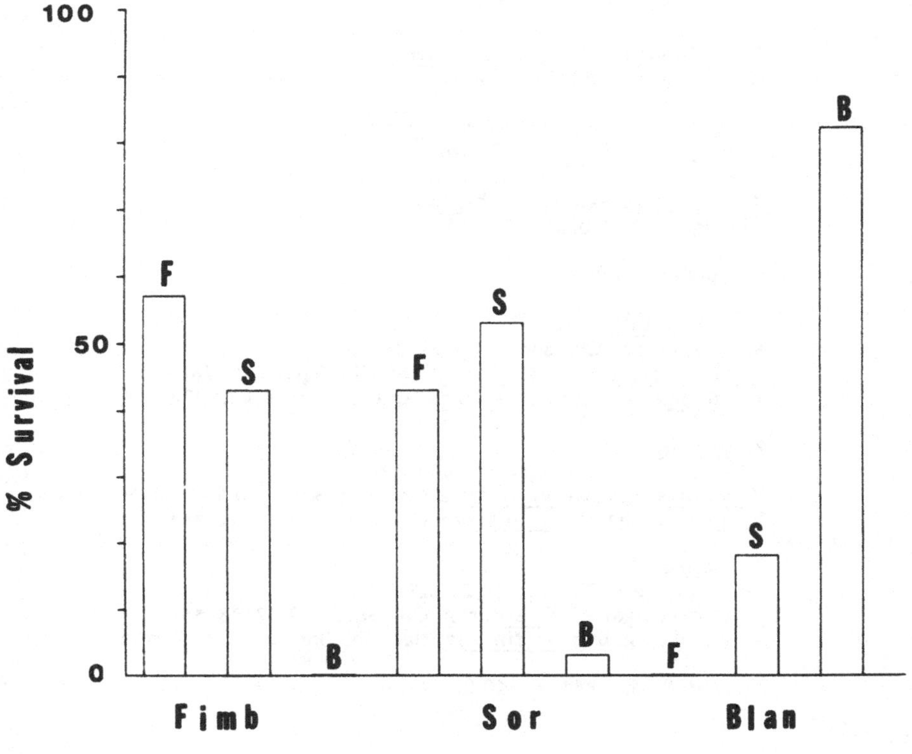

the sororia plot); V. fimbriatula was found in its original plot (where 31 out of 50 plants survived) and in the sororia plot (where 13 plants survived). Finally V. sororia was present in all three plots : 23 plants in the fimbriatula plot, 16 in the sororia plot and 3 in the blanda plot. Although more plants of V. sororia survived in the fimbriatula plot than in the sororia plot, plants of V. sororia constituted the highest percentage of the violet plants surviving in the sororia plot.

Hence the prediction is realized. V. blanda appears to be excluded from the other two plots but V. fimbriatula and V. sororia do not exclude each other. A chi-square analysis indicated a highly significant overall result (see Table 2). A plot by plot analysis also indicated highly significant results. However in a species by species comparison within each plot, the difference between V. fimbriatula and V. sororia in the fimbriatula plot was not significant (even though V. sororia was more numerous there than V. fimbriatula). It would appear then, that as predicted, V. blanda is excluded

Table 2. Chi square analysis

A. Species over all plots

1. V. fimbriatula against V. sororia
Chi Sqr for 2 d.f. = 3.0085 *

2. V. fimbriatula against V. blanda
Chi Sqr. for 2 d.f. = 48.4262 ***

3. V. sororia against V. blanda
Chi Sqr. for 2 d.f. = 32.015 ***

B. Species within plots

1. fimbriatula plot
all 3 species: Chi sqr. = 44.9653 ***
V. fimbriatula against V. sororia: Chi Sqr. = 2.57649 *
V. fimbriatula against V. blanda: Chi Sqr. = 44.9275 ***

2. sororia plot
all 3 species: Chi Sqr. = 15.75 ***
V. sororia against V. fimbriatula: Chi Sqr. = 0.437105 NS
V. sororia against V. blanda: Chi Sqr. = 15.9461 ***

3. blanda plot
all 3 species: Chi Sqr. = 21.6276 ***
V. blanda against V. sororia: Chi Sqr. = 8.57548 ***
V. blanda against V. fimbriatula: Chi Sqr. = 16.2791 ***

* = $P<0.05$; *** = $P<0.001$

from the other plots mainly due to physical limitations (inability to dissipate absorbed light energy), V. fimbriatula and V. sororia divide the habitat primarily by competitive exclusion. This conclusion can be tested experimentally, through competition experiments.

In a series of competition experiments under both high and low light conditions between V. fimbriatula and V. lanceolata (Solbrig et al., in press) it was shown that V. lanceolata outcompetes V. fimbriatula if sufficient soil water is available. Yost (1984) reached similar conclusions in an experiment in which V. fimbriatula was competed against V. sororia also under sufficient soil water availability. By "sufficient" soil water, I mean that none of the species was subjected to water stress that forced it to close its stomata to avoid leaf wilt.

SUMMARY AND CONCLUSIONS

Leaf (and other vegetative characters) are adaptive in the sense that they increase the rate of survival and reproduction of plants in certain environments. General knowledge about leaf form and function has progressed to the point that quantitative models of leaf function have been proposed and tested. In this paper such models are used to predict why four species of violets occupy preferentially certain habitats, and the prediction is tested by transplant experiments.

I wish to thank the Maria Mors Cabot Foundation of Harvard University for financial support; Elizabeth Maynard, and Ramiro Sarandon, for field and laboratory assistance; and my wife Dorothy J. Solbrig for critically reading the manuscript as well as for encouragement.

REFERENCES

Abrahamson, W.G. & Gadgil, M. (1973) Growth form and reproductive effort in goldenrods (Solidago, Compositae). Am. Nat. 107, 651-661.

Anonymous. (1962) Terminology of simple symmetrical plant shape. Taxon 11, 245, 145-156.

Antlfinger, A.E. et al. (1985) Environmental and genetic determinants of plant size in Viola sororia. Evolution, 39, 1053-1064.

Birkebak, R. & Birkebak, R. (1964) Solar radiation characteristics of tree leaves. Ecology, 45, 646-649.

Bjorkman, O. (1968) Carboxydismutase activity in shade-adapted and sun-adapted species of higher plants. Physiol. Plant., 21, 1-10.

Bjorkman, O. (1973) Comparative studies on photosynthesis in higher plants. Photophysiology, 8, 1-63.

Boardman, N.K. (1977) Comparative photosynthesis of sun and shade adapted plants. Ann. Rev. Plant Physiol., 28, 355-377.

Boyer, J.S. (1971) Non-stomatal inhibition of photosynthesis in sunflower at low leaf water potentials and high light intensities. Plant Physiol., 48, 532-536.

Brunig, E.F. (1970) Stand structure, physiognomy, and environmental factors in some lowland forests in Sarawak. Trop. Ecol., 11, 26-43.

Cook, R.E. (1979) Patterns of juvenile mortality and recruitment in plants. In: Topics in Plant Population Biology, Eds. O.T. Solbrig, S. Jain, G.B. Johnson & P.H. Raven, pp. 207-231. Columbia University Press, New York.

Cook, R.E. (1980) Germination and size-dependent mortality in Viola blanda. Oecologia, 47, 115-117.

Curtis, W.F. (1979) Adaptation in a forest floor herb in response to seasonal changes in the light environment. Ph.D. thesis, Washington University, St. Louis.

Curtis, W.F. (1984) Photosynthetic potential of sun and shade Viola species. Can. J. Bot., 62, 1273-1278.

Curtis, W.F. & Kincaid, D.T. (1984) Leaf conductance responses of Viola species from sun and shade habitats. Can. J. Bot., 62, 1268-1277.

DeCandolle, A.L.P. (1880) La Phytographie ou l'art de decrive les vegetaux consideres sous differents points de vue. Paris.

Ehleringer, J. (1976) Leaf absorptance and photosynthesis as affected by pubescence in the genus Encelia. Carnegie Inst. Wash. Yearb., 75, 413-418.

Ehleringer, J. et al. (1976) Leaf pubescence: Effects of absorptance and photosynthesis in a desert shrub. Science, 192, 376-377.

Gates, D.M. (1965) Energy, plants, and ecology. Ecology, 46, 1-13.

Gates, D.M. (1980) Biophysical Ecology. Springer, Heidelberg.

Gates, D.M. & Papian, L.E. (1971) Atlas of energy budgets of plant leaves. Academic Press, New York.

Gates, D.M. et al. (1968) Leaf temperature of desert plants. Science 159, 994-995.

Givnish, T. (1979) On the adaptive significance of leaf form. In: Topics in Plant Population Biology, Eds. O.T. Solbrig et al., pp. 375-407. Columbia University Press, New York.

Givnish, T.J. (1984) Leaf and Canopy Adaptations in Tropical Forests. In: Physiological ecology of plants of the wet tropics, Eds. E. Medina, H.A. Mooney & C. Vasquez-Janes. Junk, The Hague.

Givnish, T.J. & Vermeij, G.J. (1976) Sizes and shapes of liane leaves. Am. Nat., 110, 743-776.

Grace, J.B. & Wetzel, R.G. (1981) Habitat partitioning and competitive displacement in cattails (Typha): Experimental field studies. Am. Nat., 118 463-474.

Hale, M.G. et al. (1978) Root exudates and exudation. In: Interactions between non-pathogenic soil microorganisms and plants, Eds. Y. Dommergues & S.V. Krupa. pp. 163-204. Elsevier, Amsterdam.

Harper, J.L. & Clathworthy, J.N. (1963) The comparative biology of closely related species. VI. Analysis of the growth of Trifolium repens and T. fragiferum in pure and mixed populations. J. Expt. Bot., 14, 172-190.

Hickman, J.C. (1977) Energy allocation and niche differentiation of four co-existing annual species of Polygonum in western North America. J. Ecol., 65, 317-326.

Kincaid, D.T. & Schneider, R.B. (1983) Quantification of leaf shape with microcomputer and Fourier transform. Can. J. Bot., 61, 2333-2342.

Marshall, D.R. & Jain, S.K. (1968) Phenotypic plasticity of Avena fatua and A. barbata. Am. Nat., 102, 457-467.

Marshall, D.R. & Jain, S.K. (1969) Genetic polymorphism in natural populations of Avena fatua and A. barbata. Nature, 221, 276-278.
Mason, H.L. & Langenheim, J.H. (1957) Language analysis and the concept environment. Ecology, 38, 325-340.
Monteith, J.L. (1979) Coupling of plants to the atmosphere. In: Plants and their Atmospheric Environment, Eds. J. Grace, E.D. Ford & P.G. Jarvis, pp. 1-21. Blackwell, Oxford.
Mooney, H.A. & Gulmon, S.L. (1979) Environmental and evolutionary constraints on the photosynthetic characteristics of higher plants. In: Topics in Plant Population Biology, Eds. O.T. Solbrig et al., pp. 316-337. Columbia University Press, New York.
Mooney, H.A. & Gulmon, S.L. (1982) Constraints on Leaf structure and function in reference to herbivory. BioScience, 32, 198-206.
Mooney, H.A. et al. (1983) Physiological constraints on Plant Chemical Defenses. In: Plant Resistance to Insects, Ed. P.A. Hedin, pp. 22-36. Amer. Chem. Soc., Washington.
Mooney, H.A. et al. (1971) A mobile laboratory for gas exchange measurements. Photosynthetic, 5, 128-132.
Newell, S.J. et al., (1983) Studies on the population biology of the genus Viola. III. The demography of Viola blanda and Viola pallens. J. Ecology, 69, 977-1016.
Nobel, P.S. (1983) Biophysical Plant Physiology and Ecology. Freeman, San Francisco.
Orians, G.H. & Solbrig, O.T. (1977) A cost-income model of leaves and roots with special reference to arid and semi-arid areas. Am. Nat., 111, 677-690.
Parkhurst, D.F. & Loucks, O.L. (1972) Optimal leaf size in relation to environment. J. Ecol., 60, 505-537.
Parrish, J.A.D. & Bazzaz, F.A. (1976) Underground niche separation in successional plants. Ecology, 57, 1281-1288.
Parrish, J.A.D. & Bazzaz, F.A. (1979) Differences in pollination niche relationships in early and late successional plant communities. Ecology, 60, 597-610.
Pickett, S.T. & Bazzaz, F.A. (1976) Divergence of two coocurring successional annuals on a soil moisture gradient. Ecology, 57, 169-176.
Pickett, S.T. & Bazzaz, F.A. (1978) Organization of an assemblage of early successional species on a soil moisture gradient. Ecology, 59, 1248-1255.
Raschke, K. (1960) Heat transfer between the plant and the environment. Ann. Rev. Plant Physiol., 11, 111-126.
Raunkier, C. (1934) The Life Forms of Plants and Statistical Plant Geography. Clarendon Press, Oxford.
Sarmiento, G. (1984) The Ecology of Neotropical Savannas. Harvard University Press, Cambridge, Mass.
Sarmiento, G. et al. (1985) Adaptive strategies of woody species in tropical savannas. Biol. Rev., 60, 315-355.
Schellner, R.A. et al. (1982) Studies on the population biology of the genus Viola. IV. Spatial patterns of ramets and seedlings in three stoloniferous species. J. Ecol., 70, 273-290.
Schlichting, H. (1968) Boundary-Layer Theory. McGraw-Hill, New York.
Sharitz, R.R. & McCormick, J.F. (1973) Population dynamics of two competing annual plant species. Ecology, 54, 723-740.
Solbrig, O.T. (1981a) Studies on the population biology of the genus Viola. II. The effect of plant size on fitness in Viola sororia. Evolution, 35, 1080-1093.
Solbrig, O.T. (1981b) Energy, Information, and Plant Evolution. In:

Physiological Ecology: An Evolutionary Approach to Resource Use, Eds. C. Townsend & P. Calow, pp. 274-299. Sinauer Ass. Inc., MA. USA.

Solbrig, O.T. et al. (1980) The population biology of the genus Viola. I. The demography of Viola sororia. J. Ecol., 68, 521-546.

Solbrig, O.T. & Simpson, B.B. (1974) Components of regulation of a population of dandelions in Michigan. J. Ecol., 62, 473-486.

Solbrig, O.T. & Simpson, B.B. (1977) A garden experiment on competition between biotypes of the common dandelion (Taraxacum officinale). J. Ecol., 65, 427-430.

Taylor, S.E. (1975) Optimal Leaf Form. In: Perspectives of Biophysical Ecology, Eds. D.M. Gates & R.B. Schmerl, pp. 73-86. Springer-Verlag, New York.

Thomas, A.G. & Dale, H.M. (1976) Cohabitation of three Hieracium species in relation to the spatial heterogeneity in an old pasture. Can. J. Bot., 54, 2517-1529.

Werner, P.A. (1979) Competition and Coexistence of similar species. In: Topics in Plant Population Biology, Eds. O.T. Solbrig et al, pp. 287-310. Columbia University Press, New York.

Werner, P.A. & Platt, W.J. (1976) Ecological relationships of coocurring goldenrods (Solidago, Compositae). Am. Nat., 110, 959-971.

Yost, S.E. (1984) Habitat partitioning in Viola sororia and V. fimbriatula. PhD. thesis, City University of New York.

EVOLUTIONARY CONSTRAINTS AND SYMBIOSIS IN HYDRA

L.B. Slobodkin
K. Dunn
P. Bossert

INTRODUCTION

While there may be many conceivable solutions to the ecological and evolutionary problems faced by organisms, not all of these are equally practicable. Organisms are constrained in their structure and capacities as if there were an "economics" of somatic response and evolution (Bateson, 1963). It is impossible for any organism to find "perfect" solutions to all possible environmental problems. It is generally accepted that this impossibility has two bases. First, the ontogenetic or phylogenetic development of one adaptive property may interfere with the development of others, if only through competing demands on resources. Second, organism-environment interactions are such that different problems become more or less serious at different times and places, and this occurs in different ways for different organisms (Slobodkin, 1968, inter alia).

The existence of evolutionary constraints is generally acknowledged as a fruitful source of metaphor (Gould & Lewontin, 1979). However, it is very difficult to specify explicitly the complete set of constraints actually operating on any group of organisms. Arguments about adaptation and optimality, if they use specific examples at all, tend to focus on organisms of great complexity, in which the full context of activities and responses is unknown.

It would help if the organisms studied were sufficiently well understood and similar so that each observation could be set in the same context. For example, comparisons among dragonflies are more promising than comparisons between dragonflies and butterflies. Unfortunately, dragonflies themselves are so complex that the context of any particular anatomical or behavioural observation may not be obvious (Townsend & Calow, 1981; also comment by Hubbell on p.243 in Dingle & Hegmann, 1982).

For most taxa this situation may be irremediable. However, the freshwater polyps, the Hydra, seem so simple in behaviour and morphology and so monotonously similar from species to species as to suggest that explicit description of evolutionary constraints may be possible.

We shall describe the current state of our attempt to construct an explicit description of the somatic and evolutionary constraint system for Hydra. Initially a simple descriptive model was constructed and tested. It suggested a series of theoretical problems related to the use of resources in general and also a set of empirical questions about Hydra. Further analysis led to this simple model being replaced by a more complex theory, which is not yet complete but which has already produced questions which are of intrinsic interest.

GENERAL PROPERTIES OF EXPLICIT AND IMPLICIT RESOURCE BUDGETS

Constraints on organisms may be mechanical (D'Arcy Thompson, 1947), temporal (McFarland & Sibly, 1975) or may involve limitations in resources. We shall primarily consider constraints associated with resources. The term resource is used in a broad sense to also include energy in food.

Resource budgets may be "explicit" or "implicit". An explicit resource budget requires explicit identification and measurement of the resources themselves (i.e. energy, nitrogen, phosphorus, etc.). It is of the general form:

Resource Income = Sum of Resource Expenditures.

A resource budget for either an individual or a population is "complete" if it meets the requirements that all terms have been independently evaluated, rather than being estimated by subtraction, and that, in fact, the resource expenditure terms actually sum to the resource income, within the limits of measurement, so that there are no missing terms. It must be emphasised that completeness of the list of resource sources and sinks is essential for further discussion. The terms are assumed to be evaluated at steady state, which means that they either relate to individual adults of organisms like Hydra, planarians, or protozoans, in which senescence may not occur (Bell, 1984), or to steady-state populations.

Implicit resource budgets use a set of measurements which may be converted to resource units by suitable coefficients and ancillary measurements. One may, for example, measure food in number of prey eaten as an implicit measure of resource income, and a complete set of resource sinks, say reproduction, growth, maintenance, etc. each in their own units.

An explicit resource budget can then be written:

a_1(food) = Sum (a_2(young) + a_3(growth) + a_4(number maintained))

where the income and expenditures are measured per unit time and the a_i are suitable coefficients translating the corresponding measurements into resource units. While the evaluation of the coefficients would require ancillary measurements, certain properties of resource budgets do not depend on making the budget explicit. Only these properties will concern us here.

Optimality theory, in the sense of Townsend & Calow (1981), employs resource budgets in that any complete list of measured distinct activities constitutes an implicit set of resource sinks and any complete list of foods consumed, or of photosynthates produced, is a list of resource sources. Optimization involves apportioning of resource among the sinks.

If resource income is taken as one dimension, and each of the resource sinks be taken as a separate dimension, then the resource budget of each of a set of similar organisms, each at a particular resource income, can be represented as a single point in this multidimensional resource space.

It is to be expected that the rate of most resource expenditures will increase with resource income at low resource levels. It is not generally to be expected that this increase continues at very high resource levels. For example, for many organisms there is an attainment of some adult size (and therefore maintenance cost per individual) at some resource income level, and resource income above this level goes into reproduction rather than adult size increment (Fig. 1).

The sense of the term "monotonic surface" is that there are neither hills nor valleys on the surface. This is most readily visualized in the simple case of a resource budget containing only the two expenditure sinks "maintenance" and "reproduction" which sum to the resource income. In this case the space of interest is only three-dimensional, and the surface is two dimensional. Just this kind of surface was postulated for Hydra (see below). Such a surface is diagrammed in Fig. 2. In this simple three-dimensional case the meaning of a "monotonic surface" is that the intersection of the surface with any flat plane that passes through the entire resource income axis will be a monotonic curve.

The space in which the surface is embedded is bounded by two-dimensional planes, each having resource income and some expenditure as axes, and each passing through the origin of all other expenditure axes. If the points on the constraint surface are projected on to any of these two-dimensional planes a cloud of points is to be expected whose upper edge

Fig. 1. Possible allocation of resource which would result in a body size peak at intermediate resource levels.

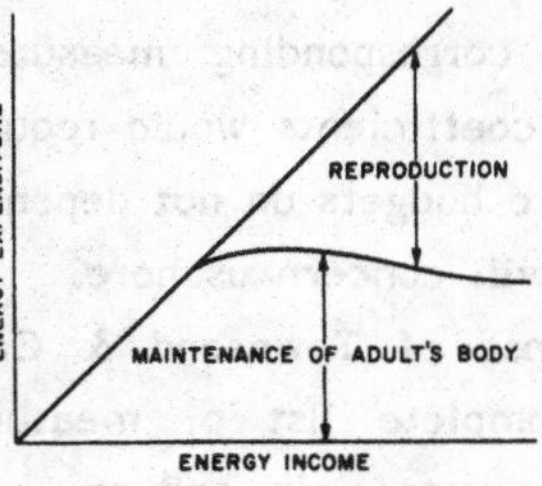

Fig. 2. An evolutionary constraint surface. I=resource income, X=body size, v=budding rate. For all Hydra there is the possibility of such a low resource income that budding does not occur. Smaller species can bud at lower resource income than larger species.

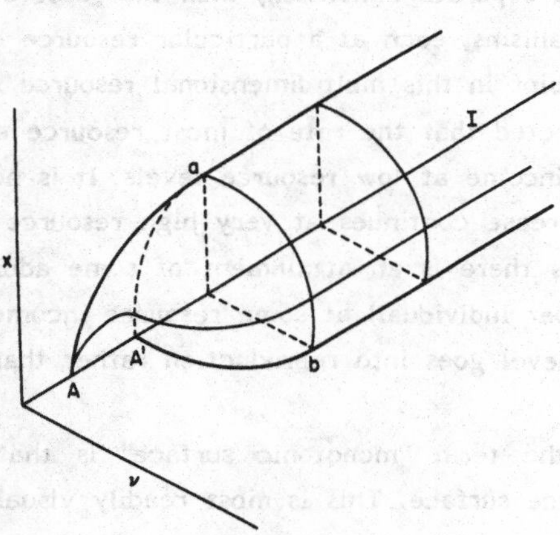

Fig. 3. Possible ways organisms might leave an evolutionary constraint surface.

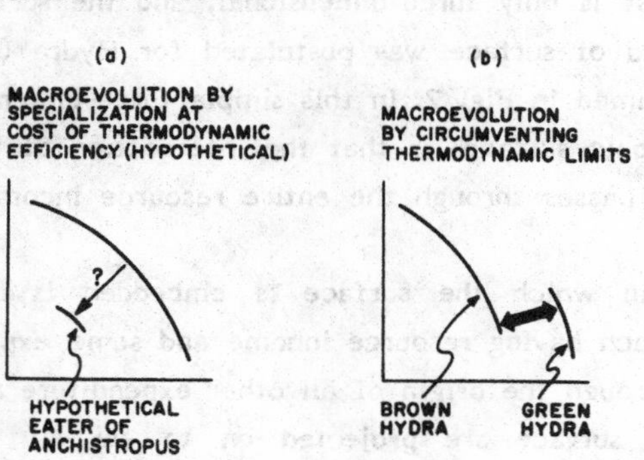

corresponds to the rim of the surface. The fact that points on a constraint surface will project onto any of the bounding planes as a cloud suggests that some of the frustrating clouds of points that are often found when ecological research workers simultaneously measure pairs of properties of related organisms, might be shadows of surfaces of higher dimension (Porter, 1976; Hutchinson, 1968; i.a.).

Organisms may be restrained from rising above the surface by efficiency considerations. Conversely, any organisms of this set that are materially less effective at resource utilization, and so might lie below the surface, may be expected to have been eliminated by competitive and selective mechanisms.

Particular species may, however, transcend the resource limitations of the other species in the clade. For example, green Hydra (individuals of the group of species usually assigned to the genus Chlorohydra) acquire resources from their photosynthetic symbionts, permitting them to move above the constraint surface (Fig. 3). It is also possible to develop particular properties that provide selective or competitive advantages which permit a species to persist even though it is below the constraint surface. For example, while Cladocera are usually food for Hydra, one Cladoceran, Anchistropus, can feed on Hydra (Borg, 1935; Hyman, 1926; Griffing, 1965). It is possible to imagine a pond in which Anchistropus is prevalent and that in this pond only a Hydra that is immune to their attack, or can feed on them will be able to survive. Transitions from the surface, such as these, may be considered as macroevolutionary events which may result in the establishment of new surfaces upon which microevolution operates.

AN EXPERIMENTAL TEST FOR THE EXISTENCE OF A SIMPLE CONSTRAINT SURFACE FOR HYDRA

It seemed possible that the relative apportionment of food resources to budding and body maintenance is the central difference among Hydra, so that they all occupy different regions of the same two dimensional surface embedded in a three dimensional space (Slobodkin, 1979). This possibility was tested for six species of Hydra (Slobodkin & Dunn, 1983).

Isolated Hydra of six species were maintained for periods of the order of thirty days. They were fed Artemia nauplii, and the number of Artemia eaten at each meal was counted, along with the buds produced. The animals were repeatedly photographed. In this way the average body size, budding rate, and feeding rate for single Hydra in isolation was measured. The measurements were all converted to rank statistics which convert any monotonic surface to a flat plane, thereby facilitating statistical tests. The cloud of points in the three-dimensional space defined by the axes 'food eaten', 'body size', and 'budding rate' was tested for agreement with two null hypotheses - that the points were randomly distributed in the three-dimensional space (i.e. were a three-dimensional array) and that they clustered around a line in the three-dimensional space (i.e. were a one-dimensional array). Both of these hypotheses could be rejected ($P < .001$). Statistically, we had demonstrated the existence of a two-dimensional surface. A more detailed account of the experiment and analysis has been presented in Slobodkin & Dunn (1983).

Our initial suggestion was confirmed but, as we shall see, in such a way as to raise important questions about the meaning of verification itself. Deeper analysis requires that the simple constraint surface diagram be seriously modified, and this modified theoretical model generates a series of implications unattainable from the simpler model.

MAP REGIONS ON THE HYDRA CONSTRAINT SURFACE

Budding rate, body size and food consumption do not completely describe the ecological state of Hydra, although they may constitute an implicit resource budget. Fortunately most other ecological properties of Hydra can be easily measured and explicitly listed. An individual Hydra may become sexual. It may float off the substrate. Sexuality and floating are reversible states, so that as circumstances change a floating animal will re-attach to the substrate and a non-sexual animal may become sexual and the converse. The initiation of floating is an active behaviour on the part of the Hydra, in response to environmental conditions, as shown by Łomnicki & Slobodkin (1966) and Slobodkin (1979).

The isolated Hydra in the Slobodkin & Dunn experiment were also examined for floating and sexuality. These properties are not randomly distributed over the surface but rather can be mapped onto specific regions (Figs. 4 and 5).

Some Hydra are green from symbiotic algae and others are brown. While the size of any Hydra depends, to some extent, on feeding rate, (and on temperature (Hecker & Slobodkin, 1976)), at any given feeding level and temperature individuals of different strains differ in size. Under any given feeding regime, green Hydra strains are smaller than brown ones. Possession of symbionts is, therefore, also confined to one region of the constraint

Fig. 4. Number of days animals were floating against body size and food consumption. Floating significantly associated with large body size and low food supply (P < .001, Cf. Slobodkin & Dunn, 1983).

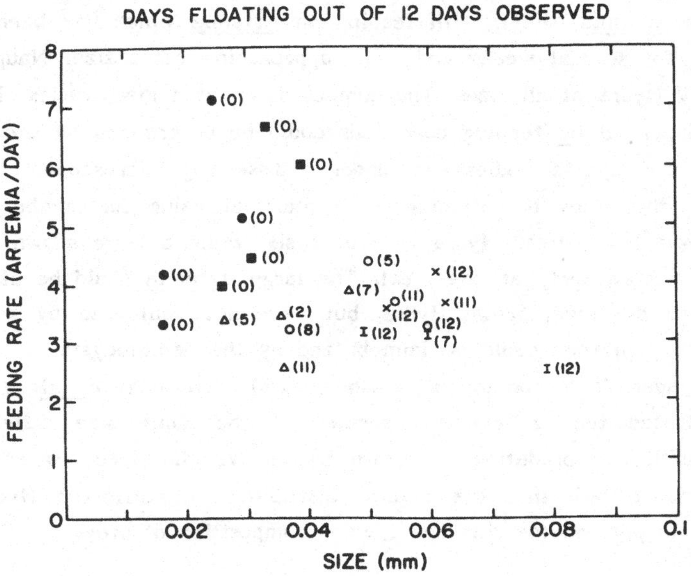

Fig. 5. Sexuality against body size and food consumption. Sexuality occurs most often at intermediate levels of food and body size. (.05 > P > .01, Cf. Slobodkin & Dunn, 1983).

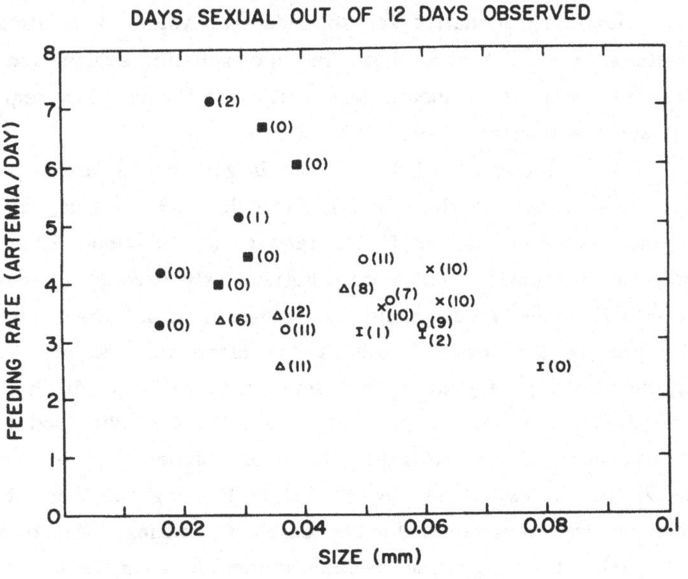

surface.

The capacity to eat foods of particular sizes and species is also delimited on the surface. This was determined by comparing feeding on newborn Artemia nauplii with feeding on Artemia which had been raised on diatoms for several weeks and were approaching adult size. Nauplii can be eaten by Hydra of all sizes. The number eaten at a meal varies. The number eaten under ad lib feeding conditions could be determined by shining a light through the Hydras' bodies and, under a dissecting microscope, counting the Artemia that they had swallowed. A maximal value for number of nauplii eaten by a small green Hydra is five or six, while a large brown Hydra may eat more than forty at one meal. The large Artemia could be attacked and swallowed by large, brown Hydra, but were not swallowed by small, green ones, although they could be immobilized by the nematocysts of small Hydra (Le Guyader & Slobodkin, unpublished data). Schwartz, et al., (1983) have demonstrated that zooplankton species of the same size differ in their susceptibility to predation by brown Hydra. We therefore suspect, but have not demonstrated, that the natural distribution of different Hydra species depends in part on the size and species composition of prey.

THE BORDERS OF MAP REGIONS

The map generates the question: What determines the borders of map regions? Explanations of the mapping exist on several levels. There are plausible evolutionary scenarios for the adaptive value of floating, sexuality and possession of symbionts, but the mechanistic explanation for these properties, and why they should be confined to particular regions of the constraint surface map are more difficult problems.

Hungry and crowded Hydra of larger species are the most likely to float off the substrate (Łomnicki & Slobodkin, 1966). This, combined with the fact that well-fed floating Hydra tend to settle (Unpublished data, L.B. Slobodkin), has the obvious value of simultaneously permitting some members of a clone to find new food supplies and of thinning out the home population, thereby improving the lot of those of the clone that do not float. Larger Hydra species float more readily than smaller ones (Slobodkin & Dunn, 1983), which is plausible from an adaptive standpoint because the food requirements for body size maintenance of larger Hydra are higher than for smaller ones. Ritte (1969) has discussed the mechanisms of floating but these did not seem to involve serious constraints. In the case of floating, therefore, we have some indication of a triggering mechanism and a plausible argument for its adaptive value, but no clear indication of why floating should be constrained to one region of the map. Is there a mechanistic explanation for why small green Hydra float so reluctantly?

Arguments about the adaptive value of sex fill entire shelves, and apply to Hydra as well as to most other organisms. Unfortunately, the mechanisms of the initiation of sexuality in Hydra are still mysterious.

A particularly interesting border on the map is that between brown and green Hydra. Symbiotic algae tend to occur in the genetically smaller Hydra species. In fact all very small Hydra are green. Symbionts provide some materials to their hosts which may permit them to withstand temporary absence of food. Small Hydra are more subject to the effects of starvation than larger Hydra and therefore there may be a greater selective value to small Hydra to harbor symbionts (Slobodkin, 1964). Muscatine & Porter (1977) have suggested that the presence of symbionts in reef corals may be an adaptation to nutrient-poor water. This type of answer is not satisfying. Saying that having algal symbionts is particularly valuable to little Hydra in no way explains why big Hydra fail to have them. Wouldn't they too enjoy some benefit? The absence of green algae as symbionts in large Hydra apparently requires explanation on some other level, possibly in terms of physiological or evolutionary constraint systems.

Bossert & Slobodkin (1983) have shown that a large species of green Hydra when semi-starved was less able to regenerate a head in the light than in darkness, while a small green species could regenerate at least as well in the light as in darkness. However, regeneration is not a normal problem in nature. Are there other ways, of greater ecological significance, in which large Hydra size and symbiosis are incompatible?

Although it has never been explicitly demonstrated, evidence is accumulating in support of a theory that the size of a Hydra is the result of the interplay of morphogenic hormones (Burnett, 1961). A sort of apical dominance is proposed and, depending upon its strength, buds are prevented from forming on the parent's body within some distance of the parent's mouth. In this way Hydra grow to a size that permits budding, thereafter shunting most new tissue growth into bud production.

Several investigators have found that at least four such morphogens exist and that two of them affect not only cell differentiation but cell division as well (Bossert & Slobodkin, in prep., Schaller, 1983). We believe that these mitogenic properties might affect the coordination of host and algal mitosis that appears to be critical to the regulation of algal density in green Hydra (McAuley, 1981, 1982). Specifically, the suite of hormones that have the morphogenic effect of creating a large body size may have mitogenic consequences that make stable endosymbiosis impossible.

Bossert & Dunn (in prep.) standardised the nutritional state of green Hydra of three strains, by feeding a given number of Artemia to each animal individually at a fixed time each day. After two weeks of this regime, the animals were fed once more and subsets were sacrificed at intervals during the twenty four hours after the last feeding. After staining, several hundred cells were examined to determine the temporal patterns of host and algal mitotic index in three strains of green Hydra in the 24 hour period following feeding. During this interval the average mitotic index of the host is approximately 1% in all strains but the average mitotic index of the algae increases from around 1% in two small strains to 3.1% in the largest strain of green Hydra.

In order to calculate growth rates from mitotic index data one must know both the number of daughter cells produced at each division as well as the duration of the scored mitotic state. As for the former, we know the Hydra mitosis results in two daughter cells but the Chlorella endosymbiont's division produces four. We have not yet measured the duration of mitosis in either the host or the algae. However, if we use values established for non-symbiotic Hydra (David & Campbell, 1972) and Chlorella (Schmidt, 1966) we find that a plausible parity in growth rate between host and algae exists in the two small strains. These same values, when applied to the large strain, yield an excessive algal growth such that the algae are apparently growing some six times as fast as the host.

In order to reconcile these data with the fact that each of these strains maintains a stable symbiosis with a relatively constant number of algae per host cell, we suggest two possible interpretations. First, it is possible that the durations of host and algal mitosis are not constant but vary in such a way that as Hydra strain size increases either host mitosis quickens or algal mitosis slows. In this way algal growth may be kept in stride with that of the host.

The other possibility is that the algae really are growing faster than the host and that this excess production is somehow culled to maintain a constant density of algae in the Hydra. In separate studies Dunn has observed temporary as well as long-term declines in algal populations in situ. These algae seem to be disappearing inside the Hydra, as they are not found in the surrounding medium. Furthermore, in other studies we have found that the growing algal population inside a well-fed Hydra has an average mitotic index lower than that of a shrinking algal population inside a poorly-fed Hydra. From this we are forced to conclude that changes in the population size of endosymbiont algae is a function of both algal birth and a significant culling process. Intracellular digestion of algae by Hydra is an obvious candidate as an in situ culling mechanism, one that we are presently investigating.

In either case the limits to green Hydra size may be defined by the limits of the Hydra's regulatory mechanisms. In the first case, large green Hydra may be near a lower limit to the duration of host-cell mitosis or near their maximum capacity to prolong algal mitosis. If the second case holds, the digestive capacity of host cells may be increasingly taxed as Hydra size increases until finally at some upper size limit, not necessarily large by non-symbiotic standards, the host can scarcely digest the excess algae as fast as they are produced. In either case the evolution of a strain of green Hydra larger than we find may be prevented by the inability of such a Hydra to prevent overgrowth by its erstwhile endosymbionts.

At the present stage of investigation, while the surface itself can be understood in terms of constraints, the mapping of properties on the surface can be partially related to a system of constraints only in the case of the distinction between small green Hydra and large brown ones.

IS THE SURFACE SIMPLE AND CONTINUOUS?

If there actually exists a single constraint surface for Hydra, it would have the implication that evolutionary processes in Hydra have been severely restricted. Hydra species would then be seen as rather minor variations on a basic theme. This is a matter of some theoretical importance.

There have been recent controversial, and even polemical, suggestions that speciation is a fundamentally different process than that described in the now classical evolutionary models of the "Modern Synthesis" of forty years ago. In fact, various authors have reintroduced the suggestion that the genetic differences found among populations of a single species are fundamentally different from the genetic differences found among different species. They also are seriously reconsidering the possibility that Goldschmidt's "hopeful monsters" (Goldschmidt, 1940) did not deserve to be so forcefully dismissed by Dobzhansky (1949) and his followers. The term "macroevolution" has gained currency. The meaning assigned to the term varies with author and context. Occasionally it refers to a change in rate of evolution (Ginzburg, 1981), or a saltation, or a change in a "Bauplan", (Gould

& Lewontin, 1979).

We would like to consider that if all Hydra, regardless of species, can be shown to lie on the same constraint surface, then macro-evolution, at least in the sense of an alteration of Bauplan can be denied. Conversely if each species of Hydra lies on a separate surface, it would permit the possibility that evolution at the species level in Hydra is fundamentally different from evolution on the level of intraspecific variation. If we had found that each species of Hydra in our experiment clustered in a distinctly different region of our three-dimensional space it might have given support to the idea of some kind of saltatory evolution. Conversely, if it had been really clear that all the experimental points were embedded in a very thin surface it would have supported purely microevolutionary mechanisms as the only ones relevant to Hydra. Unfortunately, the data did neither one. While by appropriate statistical tests we could assert that the data were spaced out as if on a surface rather than being a ball or a cigar, the data variance was much too high to tell whether the supposed surface was a single sheet, supporting only a microevolutionary interpretation, or more like a venetian blind, which would admit of macroevolution between species (Fig. 6).

Fig. 6. Difference between a clade in which only microevolution has occurred, a. macroevolution between species has occurred, b. and a clade in which macroevolution has occurred between species, but the microevolutionary events within the different species resemble each other, c.

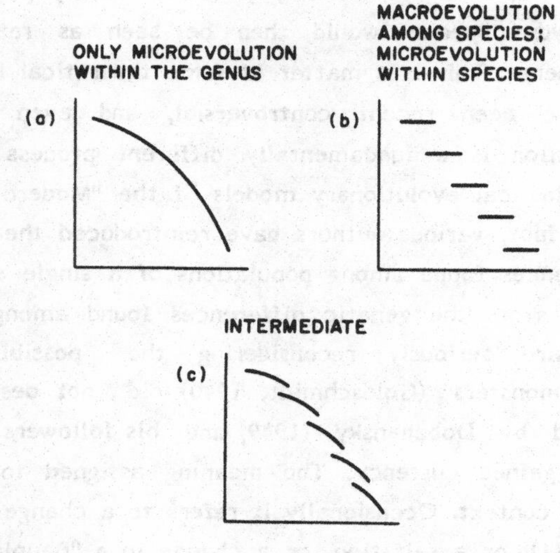

Given the expense and difficulty of the sorts of measurements required, a frontal attack on this problem, by massive increase of data, may not be the best research strategy. Reduction of variance by an increase in the precision of experimental procedures may also permit the distinction to be made. Obviously we can never completely eliminate all experimental variance. If our experiments fail to demonstrate macroevolution, it is always possible for advocates of the ubiquity of macroevolution to claim that the evidence for macroevolution is hidden in our experimental variance, no matter how small this may be. More careful measurements are of value in defining the properties of a constraint surface, and may actually demonstrate the slatted structure. However, direct falsification of the macroevolutionary hypothesis is not possible from even the most precise measurements since one might imagine discontinuities in the constraint surface hiding in even the narrowest variance cloud.

Ultimately determination of whether or not the surface is continuous depends on development of more complete models of the physiological, ecological and evolutionary properties of the clade. This model must not only unequivocally predict continuity, or discontinuity, of the surface. It must also make testable predictions which do not share the statistical ambiguities of the direct budget measurements.

Gatto et al., (in prep.) are now developing a model of Hydra ecology and evolution which predicts that the surface diagrammed in Fig. 2 must be discontinuous. It also makes testable field predictions. The model begins with a resource budget for an individual animal, equating resource income with maintenance cost and an allotment of resources to reproduction. The partitioning of resources between maintenance and reproduction can occur in accord with different possible "allocation policies". For example, either a fixed fraction of resource income is allocated to budding, or all resources above some basic income level, or some fraction above the basic level, etc. Reproductive rate is then defined as a conversion of the resource assigned to reproduction into actual buds.

The model assumes that there is no ageing in Hydra but it does develop a death rate. Density independent death in Hydra may occur because of poor water chemistry or from extreme physical conditions. Hydra may also die from predation and from starvation, which is density dependent. In the model, as in laboratory experiments (Slobodkin, 1964), Hydra populations were found to be food limited.

In addition, the model assumes that each species of Hydra is most effective at utilizing a particular size range of prey. This requirement is based on the following biological information. Very small Hydra cannot consume large zooplankton. Poorly nourished Hydra become smaller with time (Hecker & Slobodkin, 1976), which lowers the size of particle that can be swallowed. Also, very large species of Hydra have difficulty surviving and reproducing on a diet of very small zooplankters (Personal observation, L.B.S.). This may be because Hydra have no anus. A feeding Hydra gathers food particles on its tentacles and at some point must swallow them. Once the food is swallowed the Hydra become bottle-shaped and do not swallow any more until they have digested their earlier meal and regurgitated its remains. If the food supply consists of small particles available throughout the day, a large Hydra gets considerably less nourishment than if the food came as either large particles or as dense temporal aggregations of small ones.

Since the model supplies resource-dependent birth and death rates, it can produce models of population dynamics for varying resource allocation policies, food particle sizes, and temporal distributions. If we assume two competing species of Hydra, the outcome of interspecific competition will depend on their policies of resource allocation and on the food particle size distributions. While the detailed ramifications of the model are still being worked out, it is already clear that larger species of Hydra should be expected in natural water bodies which are chemically and physically salubrious and which have large, sporadically distributed zooplankters, while smaller species should be found with smaller and more evenly distributed zooplankters. The model, therefore, has two sets of predictions, namely that the constraint surface cannot be continuous and that field distribution of different Hydra species should be related to zooplankton distributions in a clear way. The first of these is difficult to test while the second may not be. The significance for ecology of being able to avoid areas of ambiguity by using formal models has been discussed elsewhere (Slobodkin, 1986).

Returning to our problem of how to decide if the constraint surface consists of a single surface or a series of surfaces, the Gatto et al., model predicts that at any given pattern of food particle availability, there will not be a single two-dimensional constraint surface for all Hydra, even in the absence of macroevolution. In the higher dimensional space in which food particle size and distribution are also dimensions there may be a single surface. The appropriate empirical test of the model consists of a study of field distribution of Hydra species and zooplankton in natural waters. This study is now in progress.

CONCLUSIONS

Dealing with Hydra, an extremely simple and relatively monotonous group of organisms, we have demonstrated that, to a first order of statistical approximation, there exists an evolutionary constraint surface derivable from a simple energy budget. The distribution of other properties among Hydra could be mapped onto this surface.

Adaptively plausible rationalisations, but no mechanistic explanation, could be provided for the fact that floating occurs in large starved non-symbiotic Hydra and that sexuality occurs under poor food conditions. It seems possible however that the confinement of algal symbionts to small Hydra species is due to a constraint system in developmental regulation which makes it impossible for large Hydra to survive with symbionts.

We considered the question of whether or not the statistical constraint surface actually consisted of a series of disjunct subsurfaces, each for a separate species. Presence of a single surface would argue for microevolutionary mechanisms being paramount in Hydra. Multiple surfaces would leave scope for macroevolution. More detailed theoretical analysis of Hydra predicts that there is most likely a separate surface for each species, given a single food type.

The analysis of whether or not there was a single surface began with a concern for whether or not macroevolutionary changes were needed to explain the properties of Hydra. At the final analysis it became clear that a broken surface in a three-dimensional space may be continuous in a space of higher dimensionality. We suggest that the evaluation of evidence for the distinction between micro- and macroevolution requires extreme care. What appears as macroevolution in one-dimensional system may not retain that appearance as more complete information becomes available. This has been suggested in different contexts by others, (Campbell, 1985; Ginzburg, 1981).

It seems likely that there is a more general procedure for determining the possibility of macroevolution having occurred in a particular clade. This is most readily seen by visualising a two dimensional case in which two properties of a group of organisms are plotted against each other. As noted by Hutchinson (1968), Porter (1976) and others this type of diagram often has a definite upper edge and is filled with a cloud of points. If the points on the upper edge all represent organisms of one subgroup of the clade, and if the points forming the filling cloud are all from other subgroups of the clade, and if addition of more descriptive dimensions resolves the cloud into a clear constraint surface, then it seems possible that the evolutionary process that made the additional dimension meaningful represents a macroevolutionary event in the sense of an alteration of Bauplan differentiating between the two clades. This is distinct from the case in which all members of a clade share a common Bauplan, in which it would be expected that addition of another dimension would reduce apparent data variance uniformly over the clade.

The research has been supported at various times by the United States National Science Foundation, NASA and the Mobil Oil Foundation. The Italian Ministry of Education and the Commission of the European Communities supported the collaboration with Gatto and Matessi.

The Istituto di Genetica Biochemica ed Evolutionisticae at Pavia and the Freshwater Biological Laboratory, Ambleside, provided working space and hospitality.

We benefitted from discussions with Scot Ferson, Lev Ginzburg and Rosina Bierbaum at Stony Brook.

REFERENCES

Bateson, A. (1963) The role of somatic change in evolution. Evolution, 7, 529-539.

Bell, G. (1984) Evolutionary and non-evolutionary theories of senescence. Am.Nat., 124, 600-603.

Borg, F. (1935) Zur Kentniss der Cladoceran-Gattung Anchistropus. Zool.Bidr. Uppsala, 15, 289-330.

Bossert, P. & Slobodkin, L.B. (1983) The effect of fast and regeneration in light vs. dark on regulation in Hydra-algal symbiosis. Biol.Bull., 164, 396-405.

Burnett, A. (1961) The growth process in Hydra. J.Exp.Zool., 146, 231-84.

Campbell, R.B. (1985) Dimension reduction projection and our perception of evolution. J.Math.Biol. (in press).

Dingle, H. & Hegmann, J.P. (1982) (Eds.) Evolution and Genetics of Life History Evolution. Springer-Verlag, New York.

David, C.N. & Campbell, R.D. (1972) Cell cycle kinetics and development of Hydra attenuata. I. Epithelial cells. J.Cell Sci., 11, 557-569.

Dobzhansky, T. (1949) Genetics and the Origin of Species, 4th Ed. Columbia University Press.

Gatto, M. et al. (In prep) Evolution with ecological and physiological constraints : Explicit predictions for Hydra.

Ginzburg, L.R. (1981) Bimodality of evolutionary rates. Paleobiology, 7, 426-429.

Goldschmidt, R. (1940) The Material Basis of Evolution. Yale University Press.

Gould, S.J. & Lewontin, R.C. (1979) The spandrels of San Marco and the Panglossian paradigm: A critique of the adaptationist programme. Proc. Roy. Soc., B 205, 581-598.

Griffing, T. (1965) Dynamics and Energetics of Brown Hydra. Ph.D. Thesis in the Department of Zoology, University of Michigan, Ann Arbor.

Hecker, B. & Slobodkin, L.B. (1976) Responses of Hydra oligactis to temperature and feeding rate. In : Coelenterate Ecology and Behaviour,Ed. G.O. Mackie, pp. 175-183. Plenum Press, New York.

Hutchinson, G.E. (1968) When are species necessary? In: Population Biology and Evolution, Ed. R.C. Lewontin, pp. 177-186. Syracuse University Press, N.Y.

Hyman, L. (1926) Note on the destruction of Hydra by a Chydorid Cladoceran, Anchistropus minor Birge. Trans. Am. Microsc. Soc., 45, 298-301.

Łomnicki, A. & Slobodkin, L.B. (1966) Floating in Hydra littoralis. Ecology, 47, 881-889.

McAuley, P.J. (1981) Control of cell division of the intracellular Chlorella symbionts of green Hydra. J.Cell Sci., 47, 197-206.

McAuley, P.J. (1982) Temporal relationships of host cell and algal mitosis in the green Hydra symbiosis. J.Cell Sci., 58, 423-431.

McFarland, D.J. & Sibly, R.M. (1975) The behavioural final common path. Phil.Trans.Roy.Soc., B 270, 265-293.

Muscatine, L. & Porter, J.W. (1977) Reef corals : Mutualistic symbioses adapted to nutrient-poor environments. BioSciences, 27, 454-460.

Porter, J.W. (1976) Autotrophy, heterotrophy, and resource partitioning in Caribbean reef-building corals. Am.Nat., 110, 731-742.

Ritte, U. (1969) Floating and Sexuality in Laboratory Populations of Hydra littoralis. Ph.D. Thesis, Univ. of Michigan.

Schaller, H.C. (1983) Hormonal regulation of regeneration in Hydra. In: Current Methods in Cellular Neurobiology Vol. IV: Model Systems, Eds. J.L. Baker & J.F. McKelvey, pp. 1-14. Wiley & Sons, New York.

Schmidt, R.R. (1966) Intracellular control of enzyme synthesis and activity during synchronous growth of Chlorella. In: Cell Synchrony-Studies in Biosynthetic Regulation, Eds. I.L. Cameron and G.M. Padilla, pp. 189-235. Academic Press, New York.

Schwartz, S.S. et al., (1983) The feeding ecology of Hydra and possible implication in the structuring of pond zooplankton communities. Biol. Bull., 164, 136-142.

Slobodkin, L.B. (1964) Experimental populations of Hydrida. In: British Ecological Society Jubilee Symposium, Eds. A. Macfadyen and P.J. Newbold. J.Anim.Ecol., 33 (Suppl.), 131-148.

Slobodkin, L.B. (1968) Toward a predictive theory of evolution. In: Population Biology and Evolution, Ed. R. Lewontin, pp. 187-205. Syracuse Univ. Press, N.Y.

Slobodkin, L.B. (1979) Problems of ecological description: The adaptive response surface of Hydra. In: Biological and Mathematical Aspects of Population Dynamics, Ed. R. de Bernardi. Mem. Inst. Idrobiol. Suppl. 37, 75-93.

Slobodkin, L.B. (1986) How to be objective in community studies. In: Neutral Models in Evolutionary Biology; Thomas J.M. Schopf Memorial Symposium. Ed. M.H. Nitecki (In press).

Slobodkin, L.B. & Dunn, K.W. (1983) On the evolutionary constraint surface of Hydra. Biol. Bull., 165, 305-320.

Thompson, D.W. (1947) On Growth and Form. Macmillan, New York.

Townsend, C.R. & Calow, P. (1981) (Eds.) Physiological Ecology: An Evolutionary Approach to Resource Use. Sinauer Associates.

GENETIC ASPECTS OF PHYSIOLOGICAL ADAPTATION IN BIVALVE MOLLUSCS

B.L. Bayne

INTRODUCTION

Physiological ecology is concerned with understanding the effects of the environment on the organisms' phenotype and with the analysis of the environment/organism interaction in terms of physiological responses which might serve as indices of Darwinian fitness. Increasingly in recent years these responses have been analysed as the acquisition of resources (usually in energy units, occasionally as nutrients) and their allocation within the organism amongst maintenance, growth, storage and reproduction. Variability in these responses has always been recognised, of course, and many attempts made to partition this variability between environmentally-induced (reversible) and non-reversible components, accepting that some elements of phenotypic expression may be due to variability in the genotype (Prosser, 1955).

However, few attempts have been made to explore the mechanisms by which genotypic variability may be expressed in phenotypic terms and, in particular, how natural selection may act on this expression to produce the physiological variability that is observed in nature. The extent of genetic diversity in natural populations is now well recognised (Nevo, 1983); the challenge is to relate this genetic differentiation to physiological differences between individuals and to features of the environment that are potentially selective. As Berry (1985) puts it: "The key to evolutionary ecology is not simple population dynamics or gene frequency change but a complementation of sensible genetic processes (involving appropriate ecological variables), with an appreciation of the effects of stressful conditions on different phenotypes".

Marine molluscs have proved a popular group of organisms for research of this kind. They are attractive to the population geneticist by virtue of their abundance, their ease of capture and, in general, their occurrence over a wide diversity of environmental conditions which may vary on both micro- and macro-geographical scales. In recent years, through the activities of aquaculture, the potential for breeding experiments with bivalves has increased. To the physiological ecologist the sessile mode of life of many species is an added attraction, with its concomitant width of physiological expression in different environments. In this contribution I will attempt a brief review of three topics that have recently attracted the attention of both physiologists and geneticists. In the first of these, an analysis of the genetic, biochemical and physiological attributes of a single polymorphic locus has contributed to our understanding of how selection operates in nature. In the second topic, multi-locus effects on the physiological processes of maintenance and growth are considered although, in this instance, causative links between genotypic and phenotypic features have not yet been drawn. Finally, one physiological process (feeding) is discussed in order to suggest the balance that exists between genotypically- and environmentally-mediated variation in a single phenotypic trait.

SELECTION AT THE Lap LOCUS IN MARINE MUSSELS

The population genetics and the biochemistry of the Lap locus in the common mussel Mytilus edulis have been reviewed by Koehn (1983). In brief, allele frequencies (there are three common alleles designated Lap^{98}, Lap^{96} and Lap^{94}) in mussels within Long Island Sound, New York, change within a single generation in a manner consistent both spatially and temporally between years. The result (Hilbish, 1985) is a persistent cline with the frequency of the Lap^{94} allele declining from approximately 0.55 at the mouth of Long Island Sound to approximately 0.15 within the Sound. Similar clines in allele frequencies at this locus which are correlated with salinity occur in other estuaries to the south of Cape Cod, USA and in the Baltic (Theisen, 1978) and Nova Scotia, Canada (Gartner-Kepkay et al., 1983). The data suggest that natural selection may be acting at this locus, the Lap^{94} gene being dominant (Hilbish & Koehn, 1985a).

In order to substantiate this claim for selection operating at this locus, however, three further requirements must be met (Hilbish & Koehn, 1985b); the different allozymes of Lap must have different biochemical properties which must, in turn, result in different physiological phenotypes and these different phenotypes must have different fitnesses under ecologically relevant conditions. Studies by Koehn and his colleagues (Young et al., 1979; Koehn & Siebenaller, 1981) have characterised the enzyme product of the Lap locus (the enzyme is aminopeptidase-I) and shown the different allozymes to differ in their catalytic efficiencies (i.e. activity per unit of enzyme concentration). The physiological expression of these allozymic differences has been investigated by Deaton et al., (1984) and Hilbish & Koehn (1985b).

Fig. 1 is taken from Deaton et al., (1984) and shows genotype-specific rates of excretion of primary amines (Fig. 1a) and ammonia (Fig. 1b). The mussels had previvously been held at 30°/oo salinity for 70h. On transfer to a lower salinity (15°/oo) excretion rates increased rapidly to take maximum values at 6h, then declining to equilibrium rates by 48h. Excretion rates for both end-products were higher in mussels with the Lap^{94} allele following transfer to low salinity; for example, for ammonia excretion these differences were statistically significant at 6h and 24h. There was no evidence of different rates of excretion at the higher salinity (i.e. time 0 in Fig. 1) or after 48h at low salinity. These results demonstrated a correspondence between a physiological process (nitrogen excretion) and a genetic difference (the presence or absence of Lap^{94}) and suggested salinity change as an environmental variable capable of evoking genotype-specific physiological rates in nature.

Fig. 1. Rates of excretion (μmoles g-1 h-1) of primary amines and ammonia by Mytilus edulis transferred from 30°/oo to 15°/oo salinity. Open circles are individuals with Lap^{94} and solid circles are individuals without Lap^{94}. Vertical lines represent standard errors. (After Deaton et al., 1984).

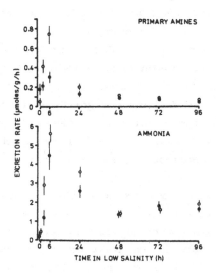

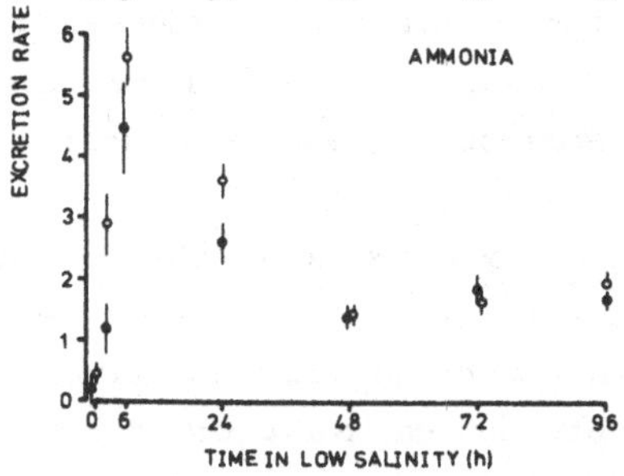

In a detailed study on mussels from the mid-point of the Long Island Sound cline Hilbish & Koehn (1985b) sought evidence of genotype-specific differences in the field and likely fitness-related consequences. Small individuals (15-20mm shell length), representing the size-class most sensitive to apparent selection against the $\underline{Lap}^{94}$ allele, were sampled frequently over 13 months and nitrogen and carbon budgets evaluated. Significant genotype-specific differences were observed; in particular, during October and November individuals with the $\underline{Lap}^{94}$ allele showed higher rates of nitrogen loss than individuals without this allele. There were no significant differences between genotypes for any component of the carbon budget. Because of their higher rate of nitrogen excretion, individuals with the $\underline{Lap}^{94}$ allele had lower scope for nitrogen growth in the autumn. For all individuals, the autumn months were characterised by minimal values for growth potential in terms of both carbon and nitrogen, possibly due to reduced food quality occurring at a time of year when metabolic rates were still relatively high. During negative growth in November, protein must have comprised less than 10% of total catabolic losses for individuals without the $\underline{Lap}^{94}$ allele, compared with more than 20% for individuals with this allele.

Hilbish & Koehn (1985b) conclude that changes in environmental salinity cause differences in rates of nitrogen loss among Lap genotypes on the shore. During a period of nutritional stress (October to December) these differences are exacerbated by episodes of negative growth and result in a reduction in the fitness of the $\underline{Lap}^{94}$ genotype, which ultimately proves lethal. It is important to note that the resulting selection occurs at one time of the year only. During the summer, for example, growth is positive and differential effects of Lap genotype are minimal. Furthermore, this selection occurs only in the first year of recruitment into the adult population. In this first year, the time available for the mussels to accumulate sufficient reserves to survive the autumn period of negative growth is limited (July to September) and, by implication, barely sufficient. In subsequent years more time is available for laying down necessary reserves and the mussels are also, by virtue of their larger size, better able to survive periods of energy and nutrient shortage.

This thorough analysis provides a convincing illustration of selection acting at a single locus and is a rare example of the "mapping" of allozymic variability on to features of phenotypic fitness. It is to be anticipated, however, that most physiological traits will depend on multiple-locus effects and therefore pose more profound difficulties in explaining physiological variability in terms of particular enzyme gene-products. An example is provided by the observed correlation between heterozygosity, measured across a number of electrophoretic loci, and growth.

MULTIPLE-LOCUS HETEROZYGOSITY AND GROWTH

Investigations with a variety of plant and animal phyla have demonstrated statistically significant correlations between mean heterozygosity and growth rate (reviewed by Mitton & Grant, 1984). Observations on marine bivalves were made by Singh and Zouros (1978) and Zouros et al., (1980); they established a positive correlation between weight and mean heterozygosity within single cohorts of the oyster Crassostrea virginica one year after settlement and concluded that over-dominance in growth rate was the most plausible explanation. Fujio (1982) made similar observations on a different oyster species, C.gigas. More recently, Zouros et al., (1983) demonstrated a positive relationship between survival and heterozygosity in C.virginica and suggested that this was a consequence of faster growth. Koehn & Gaffney (1984) confirmed a positive relationship between individual heterozygosity and size for mussels, Mytilus edulis, of similar age (Fig. 2). Gaffney & Scott (1984) have suggested that where this relationship has not been observed (e.g. Beaumont et al., 1983; Adamkewicz et al., 1984) the explanation lies in non-random, background genetic factors (e.g. linkage disequilibrium), the effects of which are exaggerated by limited effective numbers of parents employed in the experimental design.

The physiological mechanism by which more heterozygous individuals achieve higher average growth rates has been investigated by

Fig. 2. Shell length v. average individual heterozygosity (means ± 95% CI) for Mytilus edulis from Stony Brook, U.S.A. Nos. in parentheses are sample sizes. (After Koehn & Gaffney, 1984).

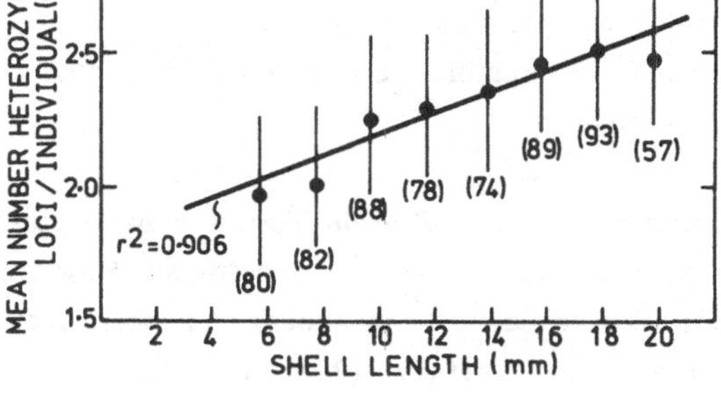

analysis of energy budgets. Koehn & Shumway (1982) recorded a strong negative correlation between heterozygosity and the rate of oxygen consumption in starved C.virginica and Rodhouse & Gaffney (1984) observed reduced rates of weight loss during starvation by the more heterozygous individuals of the same species. Garton et al., (1984) demonstrated that reduced standard metabolic rates accounted for virtually all the measured differences in growth rate amongst individual clams (Mulinia lateralis) of differing heterozygosities. Diehl et al., (1985) also observed lower rates of oxygen consumption in more heterozygous mussels (M.edulis); they concluded that reduced respiratory costs were associated more with a resistance to weight loss during starvation than with a capability for weight gain.

These studies complement earlier research on the physiological correlates of growth in strains of various species artificially selected for increased body size. For example, Medrano & Gall (1976a,b) who worked with the mealworm Tribolium castaneum, selected for high body weight at the 21-day old pupal stage. Individuals in the selected lines had metabolic rates two- to three-fold lower than in more slowly-growing individuals. Selection for large size improved efficiency of food utilization, possibly through the reduction of the energy costs of maintenance, so making more of the consumed diet available for growth. Given a relationship between rates of protein turnover and maintenance metabolism (Waterlow et al., 1978), a reduced maintenance requirement may, in turn, be due to more efficient protein turnover (e.g. Roberts, 1981).

From these various results the following hypothesis can be formulated:

1. Variations in growth rate result from higher metabolic rates in more homozygous individuals relative to individuals of greater mean heterozygosity.

2. These higher metabolic rates are due to the higher energy demands of maintenance metabolism in the more homozygous individuals and are an expression of reduced metabolic efficiencies.

3. Reduced metabolic efficiencies in turn reflect low efficiencies of protein turnover.

Growth, Energy Flux and Protein Turnover in Mytilus edulis

In order to test aspects of this hypothesis Hawkins and Bayne (unpublished data) obtained a cohort of young mussels from the same population studied by Koehn and Gaffney (1984) and Diehl et al., (1985). The mussels were divided into three groups, two comprising small individuals (8-13mm shell length, 7.6±1.1mg dry flesh weight) and one of larger individuals (20-25mm, 60.7±6.7mg). Since these animals were of the same age, the former represented slow-growing and the latter fast-growing, individuals. Measurements were then made on mussels from one of the "small" groups and from the "large" group; the second group of small individuals was grown on in the laboratory (to a size of 39.3±5.3mg dry weight) and the measurements repeated. The measurements included components of the energy budget (feeding, absorption and metabolic rates), rates of nitrogen excretion, rates of protein synthesis and breakdown (using ^{15}N techniques; Hawkins & Bayne, 1984) and individual mean multiple-locus heterozygosity as measured across five polymorphic electrophoretic loci. During the experimental period when these measurements were made, some individuals from each group were fed at a food concentration below the maintenance energy requirement, others were starved completely.

The results for the components of the energy budget, corrected for differences in body size, are in Table 1. Maintenance efficiency is defined as the efficiency with which fed mussels were able to utilize the absorbed ration to meet their demands for body maintenance, as estimated from the sum of all metabolic energy losses by individuals in the starved group. The comparison between slow-growing (low mean heterozygosity) and faster-growing (high mean heterozygosity) mussels of similar age supported the hypothesis (2 above) i.e. a higher maintenance efficiency (53.6%) in the "large" group (cf. 33.3% in the "small" group). In the third group (small individuals grown to a larger size and therefore older; low mean heterozygosity) maintenance efficiency was very low (10.8%), reflecting both genetic and age-dependent effects.

The results for protein synthesis and breakdown are in Table 2. Mussels in the faster-growing group showed higher efficiencies of protein synthesis (70.6%) than the slow-growing individuals (36.6%), as a result both of higher rates of absorption and reduced rates of breakdown. Slower-growing individuals recycled a greater proportion of the products of protein breakdown for synthesis (82.6%, cf. 53.0%), but in the two groups the estimated rates of protein loss were similar. These results point to a more intense turnover of protein in the slower-growing individuals. Hawkins (1985) calculated that protein synthesis may account for at least 16% of the standard rate of oxygen consumption; an increased rate of synthesis may therefore inflate the metabolic costs of maintenance. This is indicated in the data in Table 3,

Table 1. Energy budget for M.edulis from a single cohort. Small (7.6 mg dry wt), Large (60.7 mg dry wt) and small individuals "Grown on" for 8 weeks (39.3 mg dry wt) are compared; rates standardised to a body size of 60.7 mg by Rate(standard) = Rate(observed)$[60.7/W]^b$, where W is mean dry flesh weight and b=1.0 for metabolic rate determinations, 0.78 for excretion rates and 0.67 for absorption rates. SFG = Scope for Growth (Absorption-Metabolic losses-Excretory losses). Maintenance efficiency (see text) is calculated as follows (e.g. small individuals): [(-0.85)-(-1.02)] /0.51 = 0.333. Data from Hawkins and Bayne (unpublished). Variances and statistical significance will be discussed in a future publication but do not alter the interpretation of the data.

Animals: Condition	Metabolic + Excretory Losses: Jh^{-1}	Absorption Rate: Jh^{-1}	SFG Jh^{-1}	Maint. effic: %
Small:Fed	1.36	0.51	-0.85	33.3
Small:Starved	1.02	0	-1.02	
Large:Fed	1.41	0.70	-0.71	55.7
Large:Starved	1.10	0	-1.10	
Grown on:Fed	1.20	0.37	-0.83	10.8
Grown on:Starved	0.87	0	-0.87	

Table 2. Protein budget for M.edulis from a single cohort, standardised to a body size of 60.7 mg dry flesh wt (Table 1). EPS = efficiency of protein synthesis, calculated by: [(Synthesis-Breakdown)/Synthesis]. Recycling = proportion of breakdown that is recycled for protein synthesis, calculated by: [(Synthesis-Absorption)/Breakdown]. Protein loss is estimated from the proportion of protein breakdown that is not recycled. (Hawkins and Bayne unpublished).

Animals	Absorption	Synthesis	Breakdown	EPS %	Recycling %	Protein loss (mg per day)
	(all as mg protein per day)		>			
Small	1.54	3.23	2.04	36.8	82.8	0.35
Large	2.28	2.75	0.82	70.2	57.3	0.35
Grown on	1.16	14.26	13.42	5.9	97.6	0.32

Table 3. Protein synthesis (from Table 2) and the energy maintenance requirement for M.edulis from a single cohort, standardised to a body weight of 60.7 mg dry flesh wt (Table 1).

Animals	Rate of Protein Synthesis: mg d-1	Maintenance Energy Requirement: J d^{-1}
Small	3.23	73.4
Large	2.75	47.4
Grown on	14.26	193.1

which show the much increased maintenance energy requirements that accompany the high rates of protein synthesis in the "small" and "grown on" mussels, compared with the "large" animals. As with the energy budget, then, these results also lend support to the hypothesis (3 above) stated earlier.

Fig. 3 relates absorption to net protein balance, (i.e. protein synthesis minus protein loss) both as percent of body protein per day; there is evidence of a tight relationship, operating with a conversion efficiency of 87% (i.e. the slope of the line in Fig. 3) for all experimental groups. However, the slow- and fast-growing individuals achieved these similar conversion efficiencies through different rates (or intensities) of protein turnover, and with very different efficiencies of protein synthesis (Table 2). Three fundamental points emerge, therefore, in an integration between absorption and the energy and protein budgets. Firstly, conversion efficiencies for protein are high irrespective of rates of growth. Secondly, this constancy is achieved at different rates of protein turnover in slow- and fast-growing individuals. Thirdly, high protein turnover within the more homozygous (and slow-growing) individuals carries with it an energetic cost which is reflected in reduced maintenance efficiencies.

Another feature in the results of this experiment concerns the rates of feeding. When calculated for a standard body weight, the larger individuals had higher rates of food absorption than the smaller mussels (Tables 1 and 2). These differences were due to differences in both ingestion

Fig. 3. Net protein balance in Mytilus edulis related to absorption, both calculated as percent of body protein per day. All mussels were from a single cohort. ▲, Small size group; ●, Large size group; ■, Small individuals grown on for 8 weeks. The line is fitted as : Net balance = 0.87 Absorption -0.40

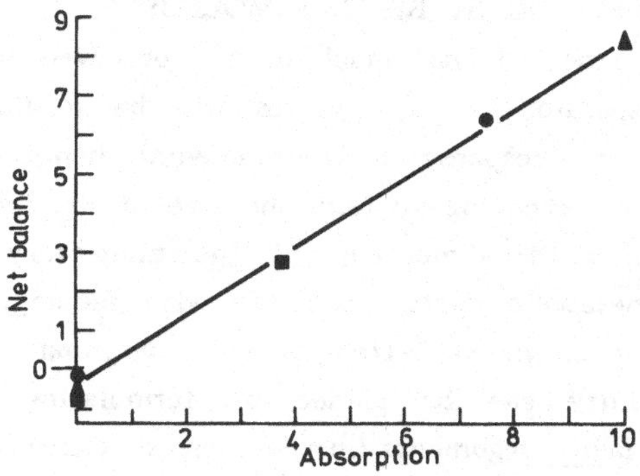

rates and absorption efficiencies (Hawkins & Bayne, unpublished data). A similar observation was made by Garton et al., (1984) on the clam Mulinia lateralis. Bivalve molluscs show considerable variability in their feeding rates (Bayne & Newell, 1983). Differences in feeding rate between individuals of different mean heterozygosity may be secondary to the primary physiological properties of the heterozygosity/growth relationship i.e. to differences in maintenance efficiency and the consequent metabolic costs. If a higher level of heterozygosity results in a slower rate of protein turnover and in a reduced maintenance requirement, then more resources can be made available for feeding as well as for growth.

These experiments are beginning to identify the physiological basis, in marine molluscs, for higher growth rates in more heterozygous individuals. The biochemical features, however, and the causal relationships, if any, between the enzymes used to score "mean heterozygosity" and differences in maintenance efficiency, remain to be elucidated. Mitton & Grant (1984) argue that the metabolic pathways normally sampled in electrophoretic analysis are likely to exert some direct influence on variability in growth rates (see also Koehn & Gaffney, 1984) and that the advantage to the individual of greater heterozygosity will likely increase with environmental heterogeneity. Future biochemical and physiological investigations can be guided by these suggestions, by the predictions of Berger (1976) on the molecular basis for heterosis, and by the conclusion reached by Watt (1985) that " ... the position of a variable locus in metabolism, with respect to the fraction of total energy flux passing through it, should scale to the evolutionary impact of functional differences ... among polymorphic genotypes at that locus".

RATES OF FEEDING BY BIVALVE MOLLUSCS

It is to be expected that much of the observed variability in rates of feeding by suspension-feeding bivalves will be mediated by the environment. Nevertheless, the responses to environmental change occur within constraints which are both morphological (e.g. the size of the ctenidia which drive the feeding currents; the dimensions of the alimentary tract) and physiological (e.g. the metabolic costs associated with feeding; profiles of activity for the digestive enzymes). Attempts to understand the adaptive features of this variability can be guided by formulating quantitative expectations, based on a priori arguments (in the sense of Calow & Townsend, 1981) and couched in the context of what is already known of the processes involved. If the hypothetical expectations are realised, they may then help to discern the "fixed" from the more "flexible" phenotypic properties of feeding behaviour and direct attention to likely heritable features.

Optimization models have proved helpful in interpreting feeding behaviour (reviewed by Pyke, 1984), including aspects of particle selection by filter-feeders (Lehman, 1976) and deposit-feeders (Taghon et al., 1978). Taghon (1981) considered relationships between feeding rate and food value in deposit-feeders and other microphages (for a more extended recent treatment see Philips, 1984). There are two basic assumptions in these models, which are repeated here: (1) individual fitness increases with net rate of energy gain; (2) animals adopt rates of feeding that maximize this net rate of gain.

Taghon's (1981) formulation may be written as follows:

$$E_n = (C.A_e.E_f) - k_1.C^x$$

where E_n is the net rate of energy gain; C is consumption (ingestion) rate; A_e is absorption efficiency; E_f is the nutritional value of the food; and k_1 and x are parameters which relate the rate of consumption to the associated energy costs. A slightly modified version of this model (Fig. 4) has been used (Bayne et al., 1984 and unpublished) to explore the feeding behaviour of marine bivalves in response to changes in food quantity and quality.

Absorption efficiency (A_e) is related to the retention time for food particles within the gut (T), weighted by the organic fraction in the available particulate matter (Fig. 4b). Gut retention time varies negatively with feeding rate (CR, litres of water cleared of food particles per hour; Fig. 4a). The metabolic rate (R) is related to consumption rate (C) as a positive exponential (Fig. 4c) and incorporates different values for standard metabolic rate by varying the y-intercept (R at C=0). The form of these relationships, and their parameter values, are generalised from measurements made on mussels (primarily Mytilus edulis but including Choromytilus meridionalis and Perna perna). The resulting model is run to calculate E_n as a function of feeding rate and the optimal feeding rate evaluated as that which maximizes E_n. The available food can be specified in terms of three variables (Fig. 4d); the total amount available in suspension, the organic fraction and its energy content. The model does not incorporate the processes of pseudofaeces production and is only used for suspended particle concentrations of less than 7 mg l^{-1}.

Fig. 5 shows some results. The optimal feeding (= clearance) rate is predicted to rise with increased organic fraction in the food (Fig. 5a), from <2 lh^{-1} for food particles of very low organic content to >4 lh^{-1} for a pure

Fig. 4. Relationships used in a model of the feeding behaviour of Mytilus edulis. a: Gut retention time, T, as a function of Clearance rate, CR. b: Absorption efficient, A_e, related to Gut retention time at two values for the organic fraction of the available food (β). c: Metabolic (respiratory, R) losses and Consumption (C) at two values for standard metabolic rate. d: Expressions for Consumption, Absorption (A) and Net energy balance (E). α = Amount of food available. ε = Energy content of the food.

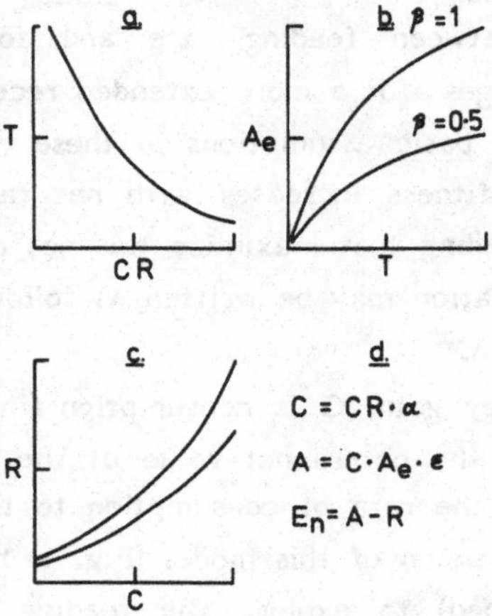

Fig. 5. Results of using model in Fig. 4 to explore possible responses to altered food quality (5a) and quantity (5b). a: Net energy gain (E_n) at diffeent CRs, with organic fraction of the food set at 0.1, 0.3, 0.6 and 1.0. For each curve optimal CR is that maximising En. Measured values for CR by Mytilus edulis at 30% (▲) and 60% (■) organic fraction are illustrated. b: Predicted optimal CRs (CR_{opt}) at different quantities of available food (seston).

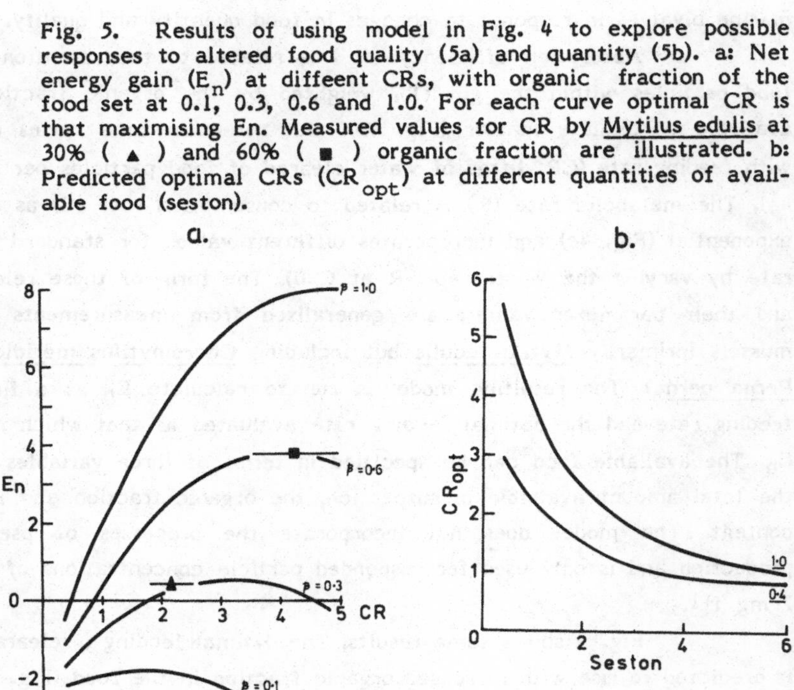

organic diet (e.g. phytoplankton). Inserted into the graph are two measurements of clearance rate taken on mussels under field conditions in which particulate organic matter was 30% and 60% of total seston; predicted (i.e. optimal) and observed feeding rates are similar (see also Bayne et al., 1984). An increase in feeding rate with food quality is in agreement with the original predictions of Taghon (1981). The model also predicts a decline in feeding rate with increased weight of particulate matter (Fig. 5b) and this too has been confirmed by among-population comparisons.

Optimal Feeding Rate in Relation to Maintenance Metabolism

In the previous section of this paper it was suggested that a relatively small reduction in standard metabolic rate would result, not only in a higher rate of net energy gain under constant dietary conditions, but also in higher rates of feeding and higher optimal clearance rates. This is illustrated in Fig. 6, a plot of results from the feeding model at three levels of standard metabolism. Due to the exponential increase in metabolic costs with increased rates of consumption, there is an enhanced increase in net energy gain per unit reduction in standard metabolic rate (by a factor of approximately x3) and an increase in the feeding rate at which E_n is maximized. In spite of a more rapid gut passage time and consequent reduction in absorption efficiency, this increase in feeding rate considerably improves net energy gain.

Variability in Feeding Rate : Constrained Optimization

Models of this type must, of necessity, consider interactions between relevant variables. If one variable acts as a constraint on another, the model may be used to explore the consequences. For example, the relationship illustrated in Fig. 4b between absorption efficiency and gut residence time depends upon assumptions regarding the rate at which digestive enzymes function during the passage of food through the gut. If the activity of the digestive enzymes is suppressed in any way, the rate at which A_e increases with T will be reduced (always assuming that gut capacity remains constant). Under these circumstances the model predicts changes in the shape of the curve of net energy gain against feeding rate (see also Taghon et al., 1978). With a rapid change of A_e with T (high digestive enzyme activity), E_n peaks sharply against feeding rate (Fig. 7); with a slow rate of change of A_e the curve for E_n is much flatter. In this latter case enzyme activity is constraining the response time for absorption and the result is a "constrained

Fig. 6. Results of using a model of feeding (Fig. 4) to explore possible relationships between E_n and CR at three levels (0.6, 0.8, 1.0) of standard metabolic rate. Optimum Clearance rates are represented as CR for which E_n is maximised; these are predicted to increase with declining metabolic rate.

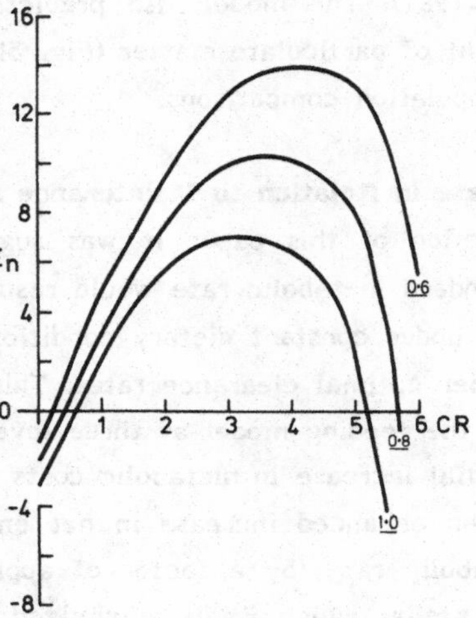

Fig. 7. Predictions from a model of feeding (Fig. 4) for E_n related to T at two levels (0.25, 1.0) for the rate of increase of absorption efficiency with increasing T (see text). The range for observed values of T for Mytilus edulis in two experiments in March and June are illustrated.

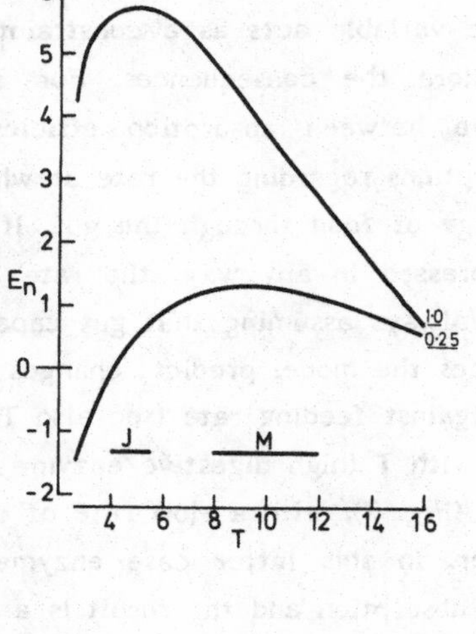

optimization" (Houston & McNamara, 1985) of feeding rate.

Hawkins et al., (1985) reported an apparently anomalous response in the feeding behaviour of M.edulis. During the winter months, individuals in the laboratory did not respond as expected to improvements in the available ration, failing to increase net energy gain; mussels in the summer responded as anticipated. Subsequent experiments have confirmed low rates of feeding and long gut residence times in the winter, compared with summer values at similar diets (Fig. 7, inserted ranges for observed values of T). The model, run with realistic constraints on absorption efficiency, suggests an explanation for reduced E_n in the winter (i.e. suppressed rates of absorption), for a failure to increase feeding rates when offered improved diets in the winter (i.e. a decline in E_n with increased rates of ingestion) and for more variability in the feeding response in winter (i.e. the flatter curve describing E_n as a function of T).

"Fixed" and "Flexible" Determinants of Feeding Rate

The above examples suggest some of the features, environmental and otherwise, that might lead to changes in rates of feeding in the field. Unfortunately, few direct and quantitative studies have been made of the reversible (i.e. flexible) and the irreversible properties of bivalve feeding rates within the full habitat range of a particular species; an indirect (and a posteriori) analysis is required.

Bivalve feeding rates, whether measured directly as rates of clearance (Vahl, 1973; Mohlenberg & Riisgard, 1979) or indirectly as rates of pumping (Meyhofer, 1985), scale isometrically with the surface area of the main feeding organs, the ctenidia i.e. feeding rate and ctenidial area increase in the same proportion with increased body size. Theisen (1977, 1982) recorded differences both in ctenidial surface area and in feeding rate for mussels (M.edulis) from Danish coastal waters. An indication of the magnitude of these differences is given in the following data (Theisen, 1982):

Station (Locality)	Ctenidial surface area (cm^2):	Mean seston conc. ($mg\ l^{-1}$):
4 (Waddensea)	18.8	>50
18 (Belt sea)	23.3	20.4
42 (Baltic sea)	31.6	2.7

The observed differences in ctenidial surface areas would imply a doubling of

Fig. 8. Clearance rates (litres per hour) related to gill (ctenidial) surface area (cm2) in Mytilus edulis from a population at Plymouth, UK. The line is plotted with a slope (exponent) of 1.0. (From unpublished data of Widdows).

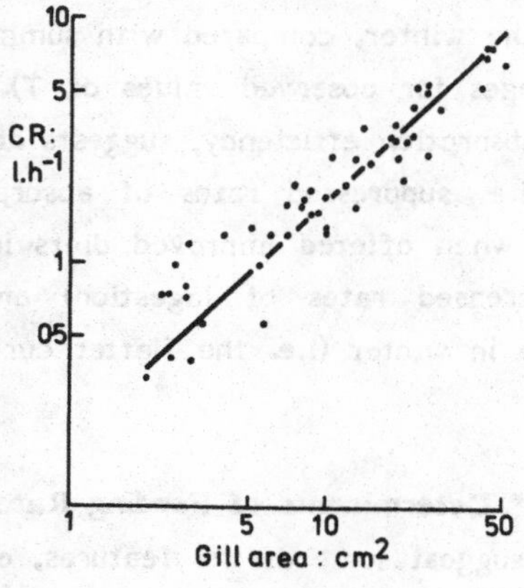

Fig. 9. Clearance rates by Mytilus edulis (litres per hour; means ± standard error). ▲ , mussels native to the Tamar;●, Swansea natives; △ , mussels transplanted from Tamar to Swansea; ○ , transplants from Swansea to Tamar. (After Widdows et al., 1984).

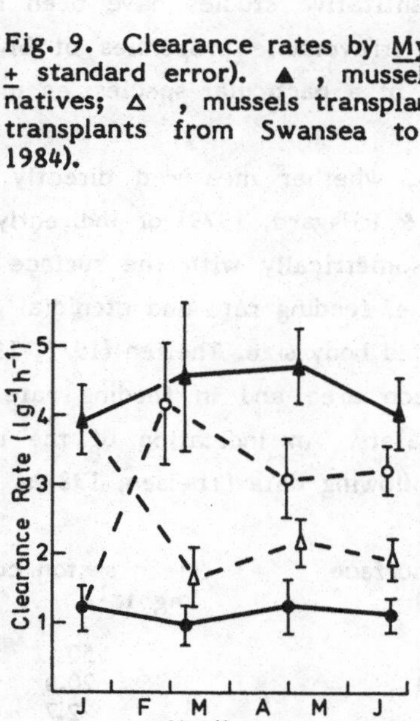

feeding rates between mussels from the Waddensea and the Baltic (see Fig. 8), correlating with average seston concentrations. Theisen (1977) had earlier measured a difference of this magnitude in the laboratory. Theisen (1978) also recorded allozyme frequency differences in mussels at these different localities and he suggests (1982) that " ... genetic causes prevail ..." in the observed morphological (and related physiological) variability between these mussels.

Widdows et al., (1984) considered the relative importance of environmental and possibly genotypic factors in determining physiological differences between two populations of mussels, by carrying out reciprocal transplants. Fig. 9 plots their results for clearance rates; physiological compensation (acclimatization) occurred, but some differences between transplanted and native mussels persisted over five months, suggesting the presence of an irreversible component in addition to a generous degree of flexibility in feeding rate.

Unfortunately, this type of experiment is unable strictly to distinguish between irreversible, non-genetic compensation and genetically-based physiological difference (Kinne, 1962). Nevertheless, Dickie et al., (1984) carried out a reciprocal three-way transplant of M.edulis and, by measuring individual growth and mortality, were able to discriminate between stock-dependent (putatively genetic) and environmental (site- and season-dependent) effects. For both growth and mortality, stock effects were considerable, giving rise, for example, to a x11.5 difference between highest and lowest rates of growth. Their results suggested a relatively low capacity for short-term physiological compensation. In particular, mussels from one stock (Ellerslie) appeared pre-adapted to maximize growth when transplanted to the other sites. These mussels had earlier been shown (Gartner-Kepkay et al., 1983) to differ in some allozyme frequencies from mussels at the other sites. As Dickie et al., (1984) conclude, " ... genetic selection for nutrition-related factors may play an important role in mussel productivity".

To unravel the balance between those physiological mechanisms which are appropriate to long-term averages in environmental states from those which are more effective as short-term responses to environmental change poses a challenge to the physiological ecologist. There is some evidence in the study of feeding and growth in bivalve molluscs of support for Bateson's (1963) notion of an "economics of somatic flexibility". Homeostatic physiological compensations provide an essential flexibility in the response to environmental change. However, these processes are "expensive" because they depend upon integrations across many inter-dependent processes and may, as a result, limit flexibility of response to further challenges. They may therefore be complemented by fixed characteristics which confer less reversible but more "economical" adaptive capacity (Bateson, 1963).

This physiological and morphological variability in bivalves occurs against a background of considerable genetic variation as measured electrophoretically (Nevo, 1983). In spite of a considerable potential for genetic mixing, due to a dispersive larval stage, micro-geographic genetic variation (over a scale of 10 - 10^3 m) may be similar in magnitude to macro-geographic variation (10 - 10^3 km). The opportunity now exists to relate this variation in space and in time to components of physiological fitness and to begin to discern how much of the total adaptive capacity of the organism has been incorporated into the genotype and how much is maintained as "somatic" flexibility.

Much of the original work reported here was carried out jointly with Tony Hawkins and with the help of Mandy Day. I am also grateful to Dick Koehn, Jerry Hilbish, Erich Gnaiger and John Widdows for many stimulating discussions and for permission to refer to their unpublished work.

REFERENCES

Adamkewicz, L. et al. (1984) Genetics of the clam Mercenaria mercenaria II. Size and genotype. Malacologia, 25, 525-533.

Bateson, G. (1963) The role of somatic change in evolution. Evolution, 17, 529-533.

Bayne, B.L. & Newell, R.C. (1983) Physiological energetics of marine molluscs. In: The Mollusca, Vol. 4, Physiology, Vol 1, Eds. A.S.M. Saleuddin & K.M. Wilbur, pp. 407-515. Academic Press, New York.

Bayne, B.L. et al. (1984) Aspects of feeding, including estimates of gut residence time, in three mytilid species (Bivalvia, Mollusca) at two contrasting sites in the Cape peninsular, South Africa. Oecologia, 64, 26-33

Beaumont, A.R. et al. (1983) Selection and heterozygosity within single families of the mussel Mytilus edulis L. Mar. Biol. Let., 4, 151-161.

Berger, E. (1976) Heterosis and the maintenance of enzyme polymorphism. Am. Nat., 110, 823-839

Berry, R.J. (1985) The processes of pattern : genetical possibilities and constraints in coevolution. Oikos, 44, 222-228.

Calow, P. & Townsend, C.R. (1981) Energetics, ecology and evolution. In: Physiological Ecology, Eds. C.R. Townsend & P. Calow, pp. 3-19. Blackwell Scientific Publications, Oxford.

Deaton, L.E. et al. (1984) Protein as a source of amino nitrogen during hyperosmotic volume regulation in the mussel Mytilus edulis. Physiol. Zool., 57, 609-619.

Dickie, L.M. et al. (1984) Influences of stock and site on growth and mortality in the blue mussel (Mytilus edulis). J. Fish. Aquat. Sci., 41, 134-140.

Diehl, W.J. et al. (1985) Physiological and genetic aspects of growth in the mussel Mytilus edulis I. Oxygen consumption, growth and weight loss. Physiol. Zool., in press.

Fujio, Y. (1982) A correlation of heterozygosity with growth rate in the Pacific oyster, Crassostrea gigas. Tohoku Jnl. Agri. Res., 33, 66-75.

Gaffney, P.M. & Scott, T.M. (1984) Genetic heterozygosity and production traits in natural and hatchery populations of bivalves. Aquaculture, 42, 289-302

Garton, D.W. et al. (1984) Multiple-locus heterozygosity and the physiological energetics of growth in the coot clam, Mulinia lateralis, from a natural population. Genetics, 108, 445-455.

Gartner-Kepkay, K.E. et al. (1983) Genetic differentiation in the face of gene flow: a study of mussel populations from a single Nova Scotian embayment. Can. J. Fish. & Aquat. Sci. 40, 443-441.

Hawkins, A.J.S. (1985) Relationships between the synthesis and breakdown of protein, dietary absorption and turnovers of nitrogen and carbon in the blue mussel, Mytilus edulis L. Oecologia, 66, 42-49.

Hawkins, A.J.S. & Bayne, B.L. (1984) Seasonal variation in the balance between physiological mechanisms of feeding and digestion in Mytilus edulis (Bivalvia: Mollusca) Mar. Biol., 82, 233-240.

Hawkins, A.J.S. et al. (1985) Feeding and resource allocation in Mytilus edulis: evidence for time-averaged optimization. Mar. Ecol. Prog. Ser., 20, 273-287.

Hilbish, T.J. (1985) Demographic and temporary structure of an allele frequency cline in the mussel Mytilus edulis. Mar. Biol., in press.

Hilbish, T.J. & Koehn, R.K. (1985a) The exclusion of a role for secondary contact in an allele frequency cline in the mussel, Mytilus edulis. Evolution, In press.

Hilbish, T.J. & Koehn, R.K. (1985b) The physiological basis of natural selection at the Lap locus. Evolution, In press.

Houston, A.I. & McNamara, J.M. (1985) The variability of behaviour and constrained optimization. J. Theor. Biol., 112, 265-273.

Kinne, O. (9162) Irreversible non-genetic adaptation. Comp. Biochem. Physiol., 5, 265-282.

Koehn, R.K. (1983) Biochemical genetics and adaptation in molluscs. In: The Mollusca, Vol. 2, Ed. P.W. Hochachka, pp 305-330. Academic Press, New York.

Koehn, R.K. & Siebenaller, J.F. (1981) Biochemical studies of aminopeptidase polymorphism in Mytilus edulis II. Dependence of reaction rate on physical factors and enzyme concentration. Biochem. Genet., 19, 1143-1162.

Koehn, R.K. & Shumway, S.E. (1982) A genetic/physiological explanation for differential growth rate among individuals of the American oyster, Crassostrea virginica (Gmelin). Mar. Biol. Lett., 3, 35-42.

Koehn, R.K. & Gaffney, P.M. (1984) Genetic heterozygosity and growth rate in Mytilus edulis. Mar. Biol., 82, 1-7.

Lehman, J.T. (1976) The filter-feeder as an optimal forager, and the predicted shapes of feeding curves. Limnol. & Oceanogr., 21, 501-516.

Medrano, J.F. & Gall, G.A.E. (1976a) Growth rate, body composition cellular growth and enzyme activities in lines of Tribolium castaneum selected for 21-day pupal weight. Genetics, 83, 379-391.

Medrano, J.F. & Gall, G.A.E. (1976b) Food consumption, feed efficiency, metabolic rate and utilization of glucose in lines of Tribolium castaneum selected for 21-day pupal weight. Genetics, 83, 393-407.

Meyhofer, E. (1985) Comparative pumping rates in suspension-feeding bivalves. Mar. Biol., 85, 137-142.

Mitton, J.B. & Grant, M.C. (1984) Associations among protein heterozygosity, growth rate and developmental homeostasis. Ann. Rev. Ecol. & Syst., 15, 479-499.

Mohlenberg, F. & Riisgard, H.U. (1979) Filtration rate, using a new indirect technique, in 13 species of suspension-feeding bivalves. Mar. Biol., 54, 143-147.

Nevo, E. (1983) Adaptive significance of protein variation. In: Protein Polymorphism: Adaptive and Taxonomic Significance, Systematics Association Special Volume No. 24, Eds. G.S. Oxford & D. Rollinson, pp. 239-282. Academic Press, London & New York.

Phillips, N.W. (1984) Compensatory intake can be consistent with an optimal foraging model. Am. Nat., 123, 867-872.

Prosser, C.L. (1955) Physiological variation in animals. Biol. Rev., 30, 229-262.

Pyke, G.H. (1984) Optimal foraging theory : a critical review. Ann. Rev. Ecol. & Syst., 15, 523-575.

Roberts, R.C. (1981) The growth of mice selected for large and small size in relation to food intake and the efficiency of conversion. Genet. Res., 38, 9-24

Rodhouse, P.G. & Gaffney, P.M. (1984) Effect of heterozygosity on metabolism during starvation in the American oyster, Crassostrea virginica. Mar. Biol., 80, 179-187.

Singh, S.M. & Zouros, E. (1978) Genetic variation associated with growth rate in the American oyster (Crassostrea virginica). Evolution, 32, 342-353.

Taghon, G.L. (1981) Beyond selection: optimal ingestion rate as a function of food value. Am. Nat., 118, 202-214.

Taghon, G.L. et al. (1978) Predicting particle selection by deposit feeders : a model and its implications. Limnol. & Oceanogr., 23, 752-759.

Theisen, B.F. (1977) Feeding rate of Mytilus edulis L. (Bivalvia) from different parts of Danish waters in water of different turbidity. Ophelia, 16, 221-232.

Theisen, B.F. (1978) Allozyme clines and evidence of strong selection in three loci in Mytilus edulis L. (Bivalvia) from Danish waters. Ophelia, 17, 135-142.

Theisen, B.F. (1982) Variation in size of gills, labial palps and adductor muscle in Mytilus edulis L. (Bivalvia) from Danish waters. Ophelia, 21, 49-63.

Vahl, O. (1973) Pumping and oxygen consumption rates of Mytilus edulis L. of different sizes. Ophelia, 12, 45-52.

Waterlow, G.C. et al. (1978) Protein Turnover in Mammalian Tissues and in the Whole Body. North-Holland Publishing, Oxford.

Watt, W.B. (1985) Bioenergetics and evolutionary genetics : opportunities for a new synthesis. Am. Nat., 125, 118-143.

Widdows, J. et al. (1984) Relative importance of environmental factors in determining physiological differences between two populations of mussels (Mytilus edulis). Mar. Ecol. Prog. Ser., 17, 33-47.

Young, J.P.W. et al. (1979) Biochemical characterization of "Lap", a polymorphic aminopeptidase from the blue mussel, Mytilus edulis. Biochem. Genet., 17, 305-323.

Zouros, E. et al. (1980) Growth rate in oysters: an overdominant phenotype and its possible explanations. Evolution, 34, 856-867

Zouros, E. et al. (1983) Post-settlement viability in the American oyster (Crassostrea virginica): an overdominant phenotype. Genet. Res., 41, 259-270.

ENERGY CONSTRAINTS AND REPRODUCTIVE TRADE-OFFS DETERMINING BODY SIZE IN FISHES

R.LeB. Dunbrack
D.M. Ware

INTRODUCTION

Fish as a group exhibit the greatest range in adult body size of any living class of organisms, spanning more than 3 orders of magnitude in length and more than 9 in weight from the goby Mistichthys luzonenis (13 mm, ~.02 g) to the whale shark Rhincodon typicus (18 m, ~22 tonnes). These extreme forms delimit a broad evolutionary potential that fishes have successfully exploited, generating in the process a remarkably diverse spectrum of adaptation. However, despite this apparent plasticity, the range of evolutionary responses available to meet the demands of aquatic life is circumscribed by an array of interacting physiological, environmental, and life-historical factors. The fitness implications of these interactions are perhaps best illustrated by the associated problems of energy acquisition and its allocation to maintenance, growth, and reproduction. For example, the rate at which energy is gained while foraging, is constrained both environmentally and physiologically. While it may respond to modifications in foraging behaviour, any behaviourally-mediated increases in foraging gain must be balanced against associated costs such as altered susceptibility to predation (Milinski & Heller, 1978; Werner et al., 1983; Dill & Fraser, 1984). After necessary maintenance costs have been paid, the remaining energy must be divided between growth and reproduction, presumably so as to maximize fitness. Investments in growth decrease size-specific mortality rate and in most cases increase future reproductive potential. Alternatively, energy invested in reproduction is clearly of greater immediate fitness value but reduces future reproductive potential (Williams, 1966; Schaffer, 1974; Charlesworth & Leon, 1976; Bell, 1984) and may increase the mortality rate (Calow, 1979; Rose & Charlesworth, 1981). The vital processes of energy acquisition, energy allocation, and mortality are thus linked in a hierarchy of constraints and trade-offs whose exact form will vary with the unique requirements of the selection environment. The purpose of this paper is to examine the body size implications of some of these constraints and trade-offs for one group of predators -- pelagic fishes.

Constraints define the limits within which trade-offs are possible, and as energy must be obtained before it can be allocated, are likely to be of particular importance to the process of energy acquisition. Thus an examination of body size constraints and trade-offs might logically begin with the relationship between body size and the way in which energy is acquired.

The pelagic organisms found in a typical North Temperate marine community range in length from 1 μm to 15 m and display a variety of techniques for prey location and capture (Fig. 1). The smallest marine creatures employ simple, nearfield structures for capturing food. The

Fig. 1. Minimum and maximum body lengths in a typical North Temperate pelagic ecosystem. Organisms less than 1 to 3 mm at birth tend to rely on tactile adaptations to locate and capture food; larger organisms tend to hunt primarily by sight; the largest fishes use a combination of visual, olfactory and acoustical searching modes (after Hardy, 1956).

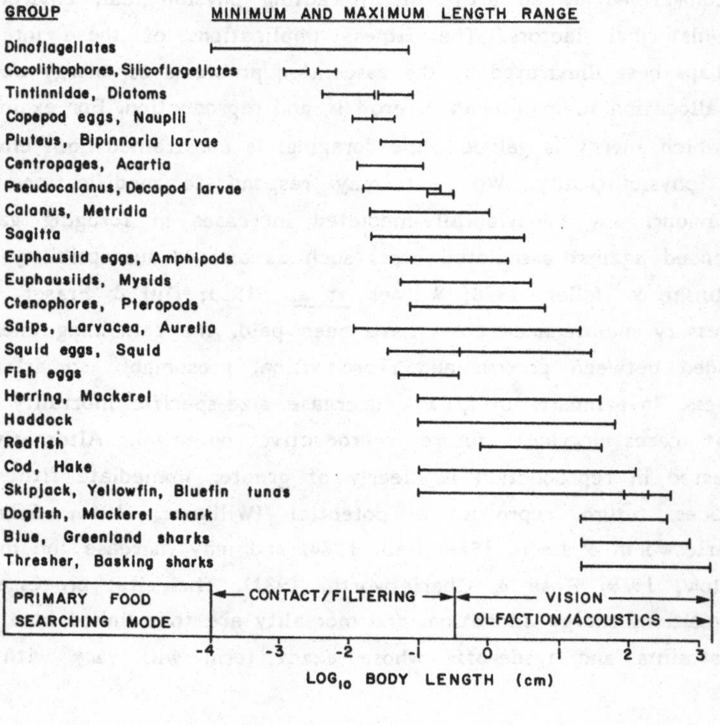

Radiolaria, for example, send out sticky protoplasmic strands that snare food particles on contact, while the Tintinnidae use cilia to create a vortex current to trap minute flagellates (Hardy, 1956). In the 1 mm size range herbivorous copepods feed by flapping their appendages and propelling water past the second maxillae, which are actively manipulated in this current to capture diatoms and flagellates (Koehl & Strickler, 1981). Predatory copepods usually detect food by sensing water disturbances via mechano-receptors on their antennae, and may leap distances greater than 1 body-length to attack motile prey (Kerfoot, 1977). In the size range occupied by pelagic fishes and their invertebrate counterparts, the cephalopods, food is located predominately by vision. The largest of the pelagic predators seem to rely on a combination of visual and non-visual modes. The lemon shark, for example, uses vision to find food at close range but relies on olfaction at distances greater than 17 m (Gilbert, 1963). Other Selachians may locate distant prey through the use of acoustic cues (Hyatt, 1979). Even the largest pelagic fish, the whale and basking sharks, which are filter feeders, may use distant sensing mechanisms to locate areas of high food concentration.

This brief survey of the foraging tactics employed by pelagic species highlights the general trend for larger body sizes to be associated with prey-location mechanisms with greater effective detection ranges. This sequence of transitions in sensory modality with body size is presumably necessitated by constraints, manifested either physiologically and/or via environmental attenuation, on the adaptive body size range of a single foraging mode. In the following sections we first consider in detail how the constraints associated with one such foraging mode -- visual prey detection -- determine the bioenergetic limits on the minimum and maximum sizes of pelagic fishes. We subsequently focus on the modification of adult body size within these limits by variations in the allocation of energy to reproduction. Finally we touch on the role of mortality in structuring the schedule of reproductive output and optimal age at first maturity.

SIZE RANGE OF VISUAL PREDATORS

Preliminaries

In this section we investigate the way in which the interaction between a visual predator and its foraging environment can lead to constraints on predator body size. Constraints are identified with reference to a standard fitness currency: surplus power (the rate at which energy gained from foraging is made available for growth and reproduction after all foraging and maintenance costs have been paid; Ware, 1980). Initially we specify the body size dependence of the variables determining foraging return. These relationships are then incorporated into a simulation model that is used to examine two food location modes, visual and non-visual. From a consideration of the characteristics of these modes we (1) delineate the body size range within which visual predation becomes the energetically more profitable tactic, and (2) explore the sensitivity of maximum adult body size, as defined by different bioenergetic criteria, to changes in foraging parameters.

We initially restrict our analyses to pelagic fishes. Trophic processes in pelagic ecosystems are characteristically body size-dependent. This and the consequent size-structure of pelagic trophic hierarchies suggests that energy constraints on body size may be of particular importance in open water environments. Pelagic systems also display two characteristics that make them especially amenable to simulation analysis:

1. Organisms occur over a wide range of body size, yet most are active particulate feeders searching for prey in a relatively homogeneous, three-dimensional world (Sheldon et al., 1972). The general features of predation thus tend to be independent of body size. Simple encounter models describing this type of predation have a notable history of success in predicting the size composition of the diets of predatory fish (e.g. Werner & Hall, 1974; Confer & Blades, 1975; Eggers, 1977; Gibson, 1980; Dunbrack & Dill, 1983).

2. The distribution of biomass with respect to the size of individual organisms can be described by a continuous function (Sheldon et al., 1972; Sheldon et al., 1973; Platt & Denman, 1978), which considerably simplifies the analysis of size-dependent trophic processes. The biomass function (Fig. 2) is defined so that the biomass density (e.g. g/cm^3) for all organisms in the size range $a \leq S \leq b$ is equal to

$$\int_a^b K1 \cdot S^{-m} ds$$

where K1 is a proportionality constant and S is a linear dimension of body size. This relationship leads to continuous expressions for both the biomass and numerical density of prey available to a predator. This in turn allows various size-dependent foraging variables to be expressed in terms of continuous functions of prey size, which can be incorporated into an easily solved expression for surplus power (see below). If predators are then assumed to vary only in body size and to be distributed according to the above density equation, a model pelagic ecosystem can be generated whose general trophic properties (e.g. body size-specific feeding rates) can be described by variations of this surplus power equation. Although this simplification lacks the realism of a multi-species approach, it allows for the investigation of ecological processes less tractable by traditional analyses (Platt & Denman, 1978).

Simulation Model

The logical basis for an investigation of the energy constraints affecting body size in pelagic fishes is the standard foraging equation (Schoener, 1971) that describes the energy obtained from foraging in terms of the components of predation. In the present context these components can be expressed as functions of predator body size to produce an equation of the form

$$Es_j = \frac{Pj\{\int_{i=a}^{b} Vj \cdot A_{ij} \cdot D_i (E_i - T_{ij} \cdot C_{ij}) di - Cs_j\}}{1 + \int_{i=a}^{b} V_j \cdot A_{ij} \cdot D_i \cdot T_{ij} \, di} - (1 - P_j) \cdot Cn_j \qquad (1)$$

where Es_j is the surplus power accumulated by a predator of size j, P_j is the proportion of the day spent foraging, a and b are the lower and upper size limits of the prey range of predator j, V_j is the velocity of search of

Fig. 2. Biomass density as a function of body size. The shaded area represents the biomass density (g/cm^3) of all particles in the size range $a \leq S \leq b$.

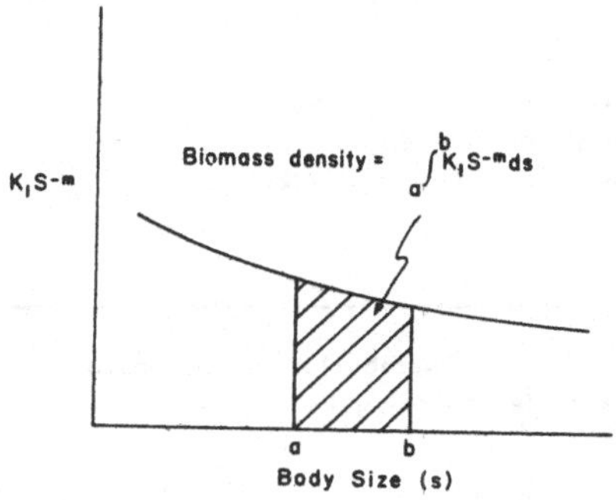

Table 1. Definitions and initial values of model parameters.

Parameter	Definition	Units	Initial value for visual predator (non-visual if different)
S_i	prey width	cm	
L_j	predator length	cm	
W_j	predator weight	g	
Es_j	surplus power	cal/day	
P_j	proportion of time spent foraging	-	0.5(1)
a	minimum prey width	cm	$0.008 \cdot L_j$
b	maximum prey width	cm	$0.05 \cdot L_j$
V_j	velocity of search	cm/s	$10. L_j^{.48}$
A_i	cross-sectional area of volume searched	cm^2	πD^2, D from Eq. 3 ($(\pi(0.25\ L_j)^2)$)
D_j	numerical prey density	num/cm^3	$K1 S_j^{-m} W^{-1}$
E_i	net energy in item of prey size i	cal	[a]750·0.7·prey wet weight (g)
T_{ij}	handling time of predator size j for prey size i	s	attack time + ingestion time + post ingestion pause
C_{ij}	handling cost of predator size j for prey size i	cal/s	attack cost = Cs_j, ingestion and post ingestion pause cost = Cn_j
Cs_j	search cost for predator size j	cal/s	[b]$1.94 \cdot 10^{-7} \cdot W_j^{0.44} \cdot V_j^{2.42}$
Cn_j	non-foraging cost for predator size j	cal/s	[b]$2.03 \cdot 10^{-4} \cdot W_j^{0.18}$
Co	inherent prey contrast	-	0.3
Cm	minimum contrast threshold	-	[c].05
α	beam attenuation coefficient	cm^{-1}	[d].001
K1	proportionality constant in biomass density function	-	$1.5 \cdot 10^{-9}$
K6	proportionality constant in eq. 3	-	$1.01 \cdot 10^{-3}$
K9	proportionality constant in ingestion time equation	-	$2.2 \cdot 10^{-10}$
E2	exponent in ingestion time eq.	-	6.62
E3	exponent in ingestion time eq.	-	0.25
m	body size exponent in biomass density function	-	1.2

[a]70% of prey energy is available for metabolism after losses due to incomplete assimilation and SDA. Prey energy is 750 cal/g (wet).
[b]Ware 1978.
[c]Hester 1968.
[d]Typical value for clear oceanic water.

predator j, A_{ij} is the cross-sectional area (which is assumed to be circular) of the volume searched by predator j for prey size i, D_i is a function of prey size i such that the numerical density of individuals in the interval $a \leqslant i \leqslant b$ is equal to $\int_{i=a}^{b} D_i \, di$, E_i is the net energy (see Table 1) obtained from an item of prey size i, T_{ij} is the handling time of predator j for prey size i, C_{ij} is the rate at which predator j expends energy while handling items of size i, Cs_j is the rate at which predator j expends energy while searching for prey, and Cn_j is the non-foraging rate of energy expenditure for predator j. We assume that fish forage so as to maximize surplus power.

The requisite surplus power function can be generated by substituting appropriate allometric expressions into (1) and solving for a number of predator sizes. The question of energy constraints on body size can then be approached by examining the form of this function and its sensitivity to changes in parameter values.

Simulation Results

Minimum size of visual predator

Fig. 1 indicates a shift to prey location modes with increased detection range, both relatively and absolutely, with increasing body size. From an evolutionary perspective we would argue that this sequence reflects relative differences in the size-specific energy return of these modes, and thus transitions from one mode to another should generally occur where the profitabilities of different modes converge. To investigate the minimum body size of visual predators, we therefore compared the size dependence of foraging return for generalised visual and non-visual predators. The body size at which these two modes become energetically equivalent, the visual/non-visual threshold, was taken as the effective minimum body size for visual predators.

Expressions and values of parameters used in the simulations are summarised in Table 1. Additional information on some of these relationships is given below.

Non-visual predator

j-Predator sizes are expressed in terms of both weight and length which we assume to be related by $W=0.005 \cdot L^{3.2}$ (Ware, 1978).

i-Prey sizes are expressed in terms of both weight and width (second largest linear dimension of body size), which we assume to be related by $W=0.862 \cdot Width^{2.8}$. Width rather than length was used for prey size as it may more closely reflect visibility and ease of capture, the latter being important in determining handling time and the upper and lower limits to prey size.

A_{ij}-As a predator moves through the water column, it searches a somewhat cylindrical volume whose longitudal axis is the predator's trajectory. The rate at which it encounters prey will be directly proportional to the effective cross-sectional area of this search volume. For both the visual and non-visual predator we assume this cross-section to be circular. For the non-visual predator we additionally assume that the radius of this area, termed reaction distance, is the same for all prey sizes and is a constant fraction of predator length. Thus $A_{ij} = \pi(K2 \cdot Lj)^2$. There are few data on the reaction distances of small non-visual predators. Giguère et al. (1982) found the reaction distances of Chaoborous larvae to be about 10% of body length. Data in Riessen et al. (1984) suggest a value near 30% for another Chaoborous species. Feigenbaum & Reeve (1977) give a value for the chaetognath Sagitta hispida which is approximately 10% of body length. For the simulations we chose an initial value of 25%.

D_i-The biomass density function (Fig. 2) can be converted to a numerical density function by dividing by prey weight. This produces the relationship $D_i = K1(S^{-m})/W$. The value of m (1.2) used in the simulations was obtained from inspection of Fig. 12 in Sheldon et al. (1972). Data in Sheldon et al. (1972) also provided an estimate of K1. They suggested that in equatorial waters the biomass density of organisms, whose weight is equivalent to that of spherical particles of diameter 100 μm, is approximately $1 \cdot 10^{-8}$ g/cm^3 per size grade. They define a size grade as the particle diameter range over which individual biomass doubles. From the relationship weight=0.862 width$^{2.8}$ (see above) it follows that a Sheldon size grade is equivalent to an increase in width by a factor of 1.28. It also follows that spherical particles of diameter 100 μm will have the equivalent biomass of fish shaped (i.e. length/width ~5) particles of approximate width .006 cm. This leads to the expression

$$10^{-8} g/cm^3 = K1 \int_{S=0.006}^{0.0078} S^{-1.2}\, dS$$

which, when solved for K1, gives a value of $1.5 \cdot 10^{-8}$. Implicit in equation 1 are the assumptions that prey are evenly distributed and that all prey encountered are also captured and consumed. Both of these assumptions are clearly unreasonable since prey may avoid visual detection in a variety of ways (e.g. by schooling or undertaking diel vertical migrations) and frequently avoid capture even if they are detected. To compensate for this effect we assume that the effective density of prey (i.e. only those detected and captured) is 10% of the actual prey density. K1 thus becomes $1.5 \cdot 10^{-9}$.

V_j-It is assumed that the velocity of search (cm/s) is a constant for any predator size. A numerical procedure which found the V_j maximizing surplus power, Es_j, for a particular predator size (the optimal velocity of search; Ware, 1978), suggested that V_j can be described by the function K8·predator length$^{0.48}$. The value of K8 obtained numerically ranged from 9 to 15 for low and high prey densities respectively. A value of 10 was used in the simulations.

T_{ij} -Predation was divided into attack, ingestion, and post-ingestion pause. Attack time was simply reaction distance/search velocity (attack and search velocities assumed equal). For a single predator size ingestion time appears to be proportional to prey size raised to the power E2 (Dunbrack, 1984). This expression was modified to extend its application to any predator-prey size combination through the inclusion of a term to account for the allometric increase in the timing of trophic processes with body size (Schmidt-Nielson, 1984; Calder, 1984). The resulting expression for ingestion time is

$$K9(\text{prey size}/\text{predator size})^{E2} W_j^{E3}.$$

Post-ingestion pause was scaled to predator size in a similar way, but was independent of prey size.

Visual predator

A_{ij} -The radius of the cross-section of the search volume of predator j for prey i (reaction distance) was set to the maximum sighting distance of predator j for prey i. This distance will vary with the size of both the predator and prey (Hester, 1968; Blaxter, 1980; Hairston et al., 1982; Dunbrack & Dill, 1983). Our reaction distance model is described below.

The visual detection of a prey item requires that the apparent contrast (Ca) of the prey exceed the contrast threshold (Ct) of the fish's visual system under the prevailing conditions. The inherent contrast (Co) of an object is commonly defined as:

$$\frac{\text{prey illumination-background illumination}}{\text{background illumination}}$$

Apparent contrast declines with distance from the observer according to

$$Ca=Co \cdot \exp(-\alpha D)$$

where α is the light attenuation coefficient and D is the distance between observer and object (i.e. predator and prey). The maximum reaction distance for any predator-prey combination will be the distance at which the apparent contrast of the prey just equals the contrast threshold of the predator. For a fixed illumination this contrast threshold will be a function of fish size (if eye size and fish size are isometric) and apparent prey size (Ware, 1973). This relationship can be conveniently approximated as

$$(2) \quad Ct=Cm+f(L_j,B)$$

where Cm is the minimum contrast threshold for fish visual systems, and f is a function of fish size (L_j) and the angle subtended at the predators eye by one of the prey dimensions (B).

We assume that the contrast threshold declines as the number of visual elements involved in the detection of a prey object increases but at a decreasing rate. This is consistent with spatial summation theories of contrast discrimination (Northmore <u>et al.</u>, 1978). The total number of visual elements involved in the detection of a particular prey object will be proportional to the absolute image size on the retina multiplied by the density of visual elements. The area of this image (A1) can be approximated by

$$A1=K3(L_j \cdot B)^2$$

where K3 is a proportionality constant. The density of visual elements (D1) will be assumed to scale as $L^v j$, so that

$$D1=K4\ L^v j$$

where K4 is a proportionality constant. The total number of visual elements is then

$$A1\ D1=K3 \cdot K4(Lj \cdot B)^2 \cdot L^v j$$

By defining the function f as

$$f=K5/(A1 \cdot D1)$$

(2) becomes

$$Ct=Cm+(K5/(A1 \cdot D1))=Cm+(K5/(K3 \cdot K4(Lj \cdot B)^2 \cdot L^v j))$$

or

$$Ct=Cm+(K6/((Lj \cdot B)^2 \cdot L^v j))$$

and the maximum sighting distance can be found by solving

$$(3) \quad Co.\exp(-\alpha D)=Cm+(K6/((Lj \cdot B)^2 \cdot L^v j))$$

numerically for D. An estimate of K6 can be obtained by rearranging (3) as

$$(4) \quad K6=(Co.\exp(-\alpha D)-Cm)(Lj \cdot B)^2 \cdot L^v j$$

and then solving for K6 directly for known or estimated values of α, D, Co, Cm, L_j, B, and v. D can then be obtained for any predator and prey size by substituting this value back into (3). This approach will be valid only for constant illumination and prey shape. The estimate of K6 used in the

Fig. 3. Visual reaction distance to the mean prey size encountered as a function of predator body size for three values of α. Boxes represent reaction distance ranges for a number of pelagic larval fishes (from Blaxter, 1980): 1. and 2. Atlantic herring, 3. plaice, 4. pilchard, 5. Coregonus spp. Arrows in upper right hand corner show the reported maximum underwater distances at which human observers can sight white netting for given attenuation coefficients (Hemmings, 1966).

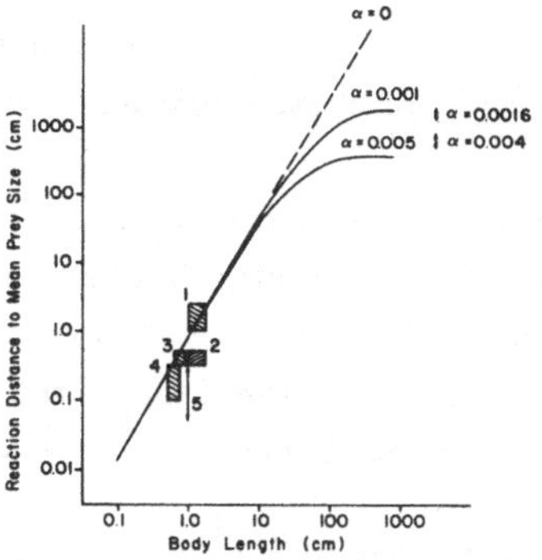

Fig. 4. Biomass encountered by visual and non-visual predators while travelling on the search trajectory (g/cm) as a function of body size. T = visual/non-visual threshold. Here the advantage shifts in favour of visual predators.

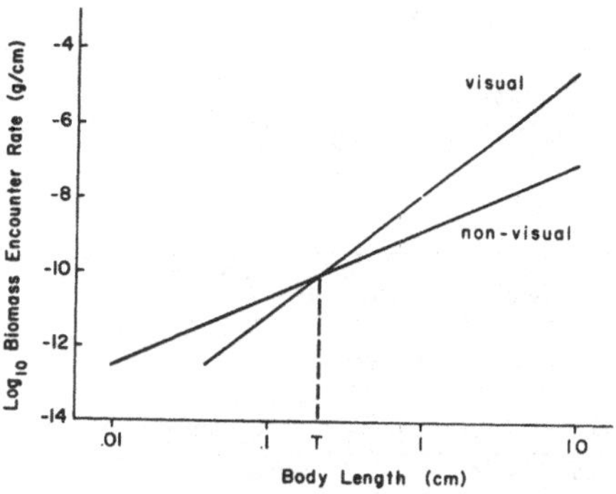

simulations was based on data in Ware (1973). This data set was chosen because it contained measurements of prey contrast that are not generally given in reaction distance studies. Other data sets from the literature for which we estimated prey contrast gave similiar values of K6.

The simulation model requires that all variables be described by continuous functions. Therefore, reaction distances for the minimum and maximum prey sizes for a particular predator size were first obtained from (3). These were then used to generate a power function relating reaction distance to prey size, which was used to calculate intermediate values. These closely approximated those calculated with (3). Reaction distances generated by the model are in general agreement with those reported in the literature (Fig. 3).

The surplus power curves for visual and non-visual predators were sufficiently sensitive to changes in the model parameters that they frequently took on negative values and were thus difficult to compare. We therefore used the biomass encountered/cm of travel on the search trajectory. This value will clearly be related to surplus power but, as it always takes on

Fig. 5. The sensitivity of the visual/non-visual threshold (cm) to changes in parameter values.

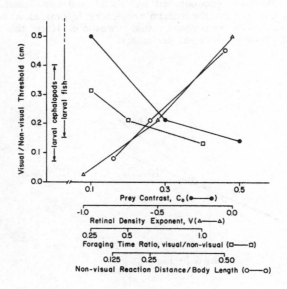

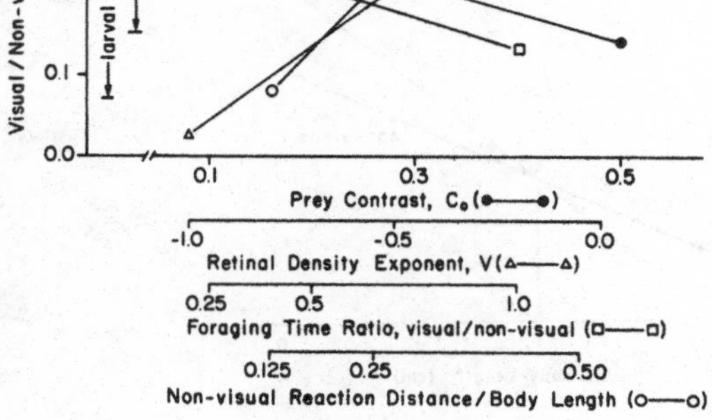

Table 2. Sensitivity of four body size criteria to changes in model parameters.

	Analysis			
	Visual/ non-visual threshold	Maximum body size	Energy maximizing body size	Time minimizing body size
A. Size with initial values (cm)	0.21	1500	800	24
B. % change in size for 10% increase in absolute value of parameters listed below:				
Attenuation coef. (α)	0	-4	-4	-1
Biomass function constant (K1)	0	4	4	-1
Biomass function exponent (m)	6	-37	-17	0
Visual element density exponent (v)	-11	0	0	-2
Prop. of day foraging (P_j)	-7*	3	3	0
Non-visual reaction distance/body length	13	-	-	-
Prey contrast (Co)	-8	8	8	-6
Contrast threshold (Cm)	0	-5	-5	-1

*The initial assumption is that the non-visual predator has twice as much time for foraging as the visual predator. The 10% increase refers to increasing the proportion of the day available to the visual predator for foraging and keeping that for the non-visual predator constant.

Fig. 6. Surplus power (cal/d) as a function of body length. The horizontal line represents a surplus power of zero, values below this represent negative surplus power. Point 1 is the absolute maximum body size, point 2 is the energy maximizing body size, point 3 is the time minimizing body size (see text).

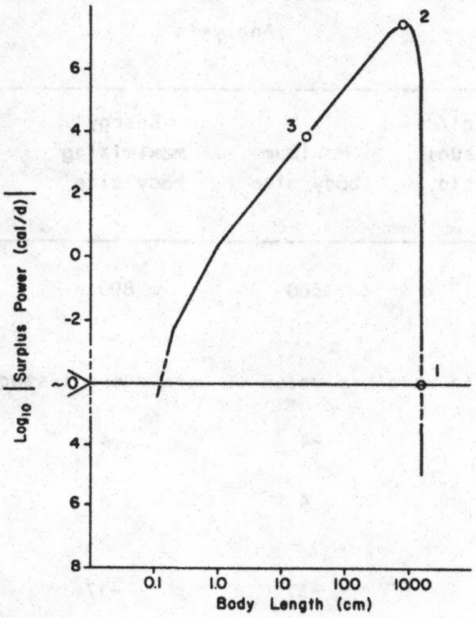

Fig. 7. The sensitivity of the energy maximizing body size (EMS) to changes in parameter values.

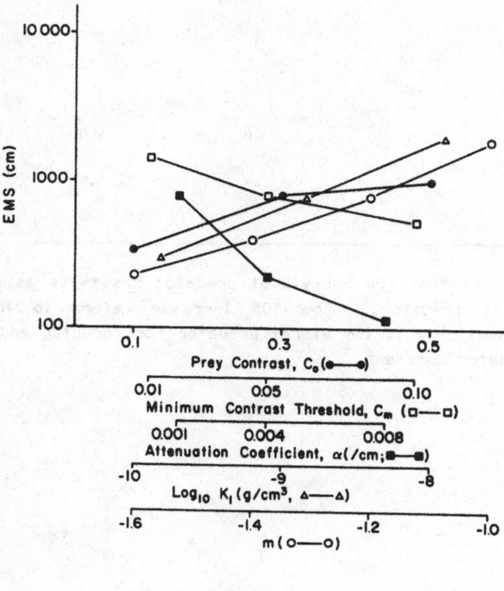

positive values, is more useful for comparative purposes.

The results using initial values (Table 1; Fig. 4) indicate that biomass encounter rates are higher for non-visual predators at small body size. At larger sizes this relationship is reversed because the reaction distance of visual predators increases more rapidly with size than does the contact radius of non-visual predators. This is illustrated in Fig. 3 which shows that at small predator sizes, reaction distance scales approximately as length$^{1.75}$ while the contact radius of non-visual predators is assumed to be isometric with predator length.

The sensitivity of the visual/non-visual threshold to changes in the various parameters of equation 1 was also examined (Table 2; Fig. 5). Prey contrast (Co), the exponent of the biomass equation (m), the body size retinal density exponent (v), and the proportion of the day available for the foraging of visual predators relative to that of contact predators (P_j) all affected the threshold size, although in general the ratio (%change in threshold/%change in parameter) was less than 1.

Maximum size of visual predators

Our approach to the question of energy constraints on maximum adult body size was somewhat different from that used for minimum size. A comparative approach is difficult because the characteristics of non-visual predation which would replace visual predation at large body sizes are poorly understood. For this reason we did not consider the upper transition from visual to non-visual prey detection directly, but instead restricted our analysis to an examination of the constraints imposed on body size solely by the surplus power characteristics of visual predation.

Using the initial parameter values (Table 1) equation 1 was solved for a number of visual predator body sizes to generate a surplus power curve, the general form of which is shown in Fig. 6. An examination of this figure suggests at least two bioenergetically relevant criteria that can be used to define maximum adult body size.

1. The body size at which surplus power is zero and the fish is in a state of energy balance with its environment (Fig. 6; point 1). Here, foraging gains just meet the cost of maintenance and no surplus is available for growth or reproduction. This is clearly the absolute maximum body size and although of interest energetically, it is not likely to be of ecological significance.

2. The body size at which surplus power is maximized. (Energy

maximizing size; Fig. 6; point 2). At this size energy should be invested primarily in reproduction since further increases in body size would reduce surplus energy in an absolute sense (Sebens, 1979). Thus the energy maximizing size should represent an effective upper size limit.

A third possible criterion identified by foraging theory (Schoener, 1969) but not evident from an examination of Fig. 6, is:

3. The body size where the foraging time required to obtain a fixed proportion of the maximum ration is minimized (time minimizing size; Fig. 6; point 3). Alternatively, if energy rather than time is limiting, this is the size at which predators will obtain the greatest proportion of their maximum ration in a fixed time. Although this size may not be directly linked to an identifiable energy constraint on maximum body size it may define the minimum size at which pelagic fish should mature (see Discussion).

The way in which point 2, the energy maximizing size, responds to parameter changes is shown in Fig. 7. The response of absolute maximum size (point 1) is essentially the same except the values scale further up the body size axis. The sensitivities of the three points to changes in the model are summarized in Table 2.

The body size which minimizes foraging time (Fig. 6; point 3) is noteworthy for two reasons. First, it is considerably smaller than points 1 and 2, and second it is better defined, being less sensitive, both relatively and absolutely, to changes in foraging parameters (Table 2).

Discussion

Our analyses of the foraging energetics of pelagic predators indicate that visual predation should become a viable tactic when body length exceeds 0.5 to 5 mm. At smaller body sizes, reaction distances and thus prey encounter rates, are predicted to fall well below those expected from mechano-sensory modes of prey detection. The extremely short reaction distances of smaller visual predators result from both small prey size and low visual acuity, the latter a consequence of the reduced light gathering ability of a smaller lens (Walls, 1963; Hester, 1968; Blaxter, 1980). The predicted range is in general agreement with the larval sizes of a number of pelagic fish species and it is noteworthy that cephalopods, particulate feeders with an eye strikingly convergent to that of vertebrates, have similar minimum sizes (Fig. 5). The agreement between predicted and observed may result from the inverse relationship between the size and number (fecundity) of eggs produced by a female. By reducing egg size to the minimum required for successful visual predation by newly hatched larvae, effective female fecundity is maximized and it is reasonable to expect that under certain circumstances this reproductive tactic will also maximize fitness.

The existence of an energy maximizing size (EMS) for pelagic fishes is largely a consequence of light attenuation in the water column and the resulting limit on prey detection range. However, the exact position of the EMS is sensitive to changes in a number of environmental variables, and we would therefore expect observed maximum body sizes to display some environmental specificity. Modifications of these maxima by selection acting on the visual system is possible but would be limited by the pervasive effect of light attenuation.

The energy maximizing sizes obtained by simulation (Table 2) are in general larger than the maximum reported sizes of pelagic fishes. This result is not unexpected as the constraints imposed by the requirements of fitness maximization are likely to differ from those limiting maximum body size (see next section). These theoretical values are therefore not likely to be good quantitative predictors of realised maximum body size, but are useful for examining the factors that limit it. From this perspective the results of the simulations are most informative when interpreted comparatively. For example, maximum body size is negatively correlated with the light attenuation coefficient and positively correlated with biomass density (Table 2). In addition, light attenuation and biomass density are themselves positively correlated in most marine environments (i.e. a high biomass density causes greater scattering and absorption of light and therefore increases the attenuation coefficient). This interaction makes it difficult to predict how the limits to maximum body size would respond to increases in prey standing stocks, such as might occur in real systems following either natural or induced enrichment. However, the simulation model can be adapted to examine such questions. A comparison of the EMS in two otherwise identical environments, one characterised by high attenuation and high standing stocks (attenuation coefficient=.004; $K1=1.2\cdot10^{-8}$; EMS=570 cm) and the other by low attenuation and low standing stocks (initial values in Table 1; EMS=800 cm) indicates that the greater prey detection distance possible in clear water (low attenuation) can more than compensate for the decrease in prey density.

The time minimizing body size (TMS) is, in general, less sensitive to changes in foraging parameters than either the maximum or minimum sizes. In addition to foraging time and relative surplus power (see above), a number of other energy related body size correlates reach maximum or minimum values close to this size. These include numerical prey encounter rate (maximized), time spent searching (minimized), reaction distance to mean prey size, as a proportion of body length (maximized), and time available for non-foraging activities including predator avoidance (maximized). The number of fitness relevant functions reaching extreme values close to the TMS suggests that it may approximate the optimal size at maturity for large numbers of pelagic species. The distribution of the maximum adult sizes of species from a number of pelagic families supports this (Fig. 8).

The simulation results suggest that selection acts on the components of predation in a body-size dependent fashion. At small sizes (close to the time minimizing size) the relative prey location abilities of visual predators are maximized and the main foraging challenge facing predatory fish is likely to be the capture of relatively easily located prey. If selection acts most strongly on the weakest link in the foraging chain, the greater evolutionary elaboration of foraging-related adaptations should be associated with prey capture rather than detection. For example, specialized mouth structure and specialization on particular prey types or time of feeding should be most evident in fish close to the TMS. With increasing body size, relative search time increases and there will be fewer opportunities for specialization. As suggested in Fig. 1 this would tend to favour the development of prey location modes with greater detection distances to augment a final visually mediated attack. Other adaptations, such as a high proportion of aerobic muscle, increased prey size range, or the development of specialised searching tactics such as group hunting might also become increasingly evident at larger body sizes due to the greater proportion of time that must be spent in active search.

To generate a single surplus power curve spanning all body sizes, we have assumed that increases in predator size are accompanied by increases in the size of prey consumed. However, we do not suggest that the same surplus power curve characterizes all pelagic fishes. For most species there will be restrictions on adaptive changes in foraging behaviour as body size increases, at some point prey size will no longer increase in proportion to predator size, and a local EMS will occur. To accommodate this reality constraints on the increase in prey size with predator size could be introduced. By doing this repeatedly, a spectrum of similar curves would be generated each corresponding to the energy constraints dictated by a particular ecological specialization (Jones & Johnson, 1977). For each of these curves the energy maximizing size would indicate either a maximum adult body size or the size at which changes in predatory habit should occur. This modification could allow the model to be applied to a broader range of questions relating to the diversity of both species and life history in pelagic fishes such as, for example, the evolution of diadromy and other size specific migrations or changes in predatory habit (e.g. Bachman, 1981).

REPRODUCTION AND MORTALITY TRADE-OFFS

In this section we consider how the actual body size attained by adult fish is modified, within the limits set by the surplus power curve, by the requirements of reproduction.

Asymptotic Body Size

An immature fish, by definition, invests its surplus power exclusively in growth. However, as it approaches sexual maturity it must begin allocating some portion of this surplus to reproduction. Between the size at first reproduction and the EMS both surplus power and reproductive output appear to be related to body size by simple power functions (Fig. 6; Ware, 1980), allowing this explicit trade-off to be expressed as

$$(5)\quad dW/dt = E - R = A \cdot W^{b} - C \cdot W^{d}$$

where dW/dt is the instantaneous growth rate, E is the surplus power, R is the reproductive output, W is body weight, and A, b, C, and d are constants. If d>b there may exist a maximum, or asymptotic, adult body size (Wmax) less than the EMS, determined solely by the surplus power and reproductive output schedules. An expression for Wmax can be obtained from (5) by setting dW/dt=0 to give

$$Wmax = (C/A)^{(1/(b-d))}$$

A limited amount of empirical information is available on the parameters of the reproduction and surplus power schedules which determine Wmax. For the marine fishes examined so far (Ware, 1980; Myers & Doyle, 1983) the surplus power weight exponent, b, varies between 0.5 and 1.1. Although b tends to be around 1 for most pelagic species, some demersal forms like American Plaice have similar values (Myers & Doyle, 1983), and Pacific herring, a decidedly pelagic species, has a value significantly less than 1 ($b \sim 0.57$; Ware, unpublished data). As yet there are no obvious generalizations concerning how surplus power scales with body size, although it has been suggested that the magnitude of b may decline with size due to the general decrease in both the biomass and numerical density of prey organisms; b may also be related to the foraging mode and size range of

Fig. 8. Frequency distribution of maximum body lengths of pelagic fishes. Data on the Engraulidae, Clupeidae, and Scombridae from Pauly (1980), and on the Osmeridae and Salmonidae from Hart (1973). In the composite panel notice that the time minimizing size (TMS) spans the observed modal size classes (15-25 cm); the >150cm category also includes the pelagic sharks listed in Hart (1973).

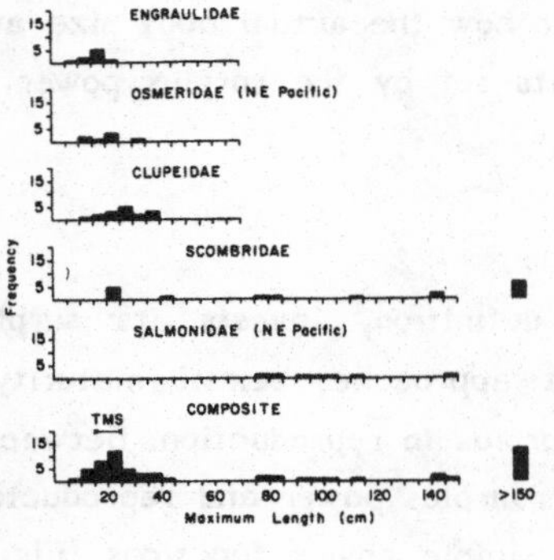

Fig. 9. Logarithmic relationship between reproductive effort (RE; dimensionless) and body weight (g), from the size at first maturity (Wm) to the asymptotic size (Wmax), in various species. Since the weight exponent in each case was very close to 0.7, this was taken to be the common slope, and the respective proportionality constants were estimated so that RE=1 at Wmax. Note that the species with the lowest reproductive effort (and lowest instantaneous annual natural mortality rates, M) tend to show the greatest post-maturational growth, Wmax-Wm (RE data from Ware, 1980, Table 3, and unpublished data; mortality rates from Pauly, 1980).

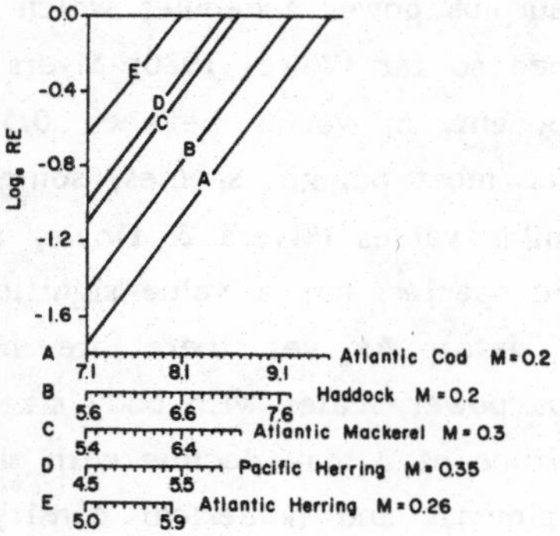

available prey (previous section; Ware, 1980). Far less is known about reproductive output, since most studies have concentrated on fecundity and have failed to determine mean egg weights or the calorific value of gonadal products. As a generalisation Roff (1983) observed that the reproductive output exponent, d, varies between 1 and 1.9.

Some indication of the sensitivity of asymptotic size (Wmax) to variations in surplus power and reproductive output schedules can be obtained using simple simulation procedures. For parameter values typical of Pacific herring populations (A=4.9, b=0.575, C=0.0852, d=1.26), we calculate that a 10% increase in A and C produces a 13% increase and 19% decrease in Wmax, respectively. All else being equal, changes in the reproductive output coefficient, C, are likely to have a slightly greater effect on Wmax. Changes in the weight exponents b and d would be expected to have qualitatively similar effects to those of their respective coefficients.

Reproductive Effort

The size dependence of the growth/reproduction trade-off can be approached more directly by expressing reproductive output as a fraction of surplus power. The resulting metric, termed reproductive effort (RE), is expressed as

$$RE=R/E=(C/A)W^{(d-b)}$$

Published data indicate that despite large differences in A, b, C, d, and the size at first maturity, the reproductive effort exponent (d-b) is remarkably constant (~.7) amongst the disparate assemblage of species that has been examined in some detail (Fig. 9). If age and size are correlated, the invariance of this life history parameter and the consequent similarity of the reproductive effort schedules of these species when normalised for size at first reproduction (Fig. 9) parallels the general constancy of a number of size-dependent reproductive processes when normalised for lifespan (i.e. in physiological time; Calder, 1984).

Size and Age at Maturity

Although asymptotic size is proximally determined by the schedules of reproduction and surplus power, these schedules must themselves ultimately reflect the constraints of a fitness maximizing selection process, involving mortality as well as energy acquisition and allocation (Jones & Johnston, 1977). A particularly important outcome of this process will be the size, Wm (or age, Tm), at first reproduction. For iteroparous species, this point will clearly have both a greater relevance to fitness and be more responsive to selection than Wmax which, because of post-maturational mortality, may rarely be attained. We will briefly consider how the optimal age at maturity (Tm,opt), i.e. that which maximizes fitness, responds to changes in life-history parameters. We focus on age rather than size at maturity as it is more usually identified in population studies and is computationally more tractable when dealing with temporal rates such as mortality.

Assuming that lifetime reproductive output (LTRO) is an appropriate measure of fitness, Tm,opt can be obtained by maximizing the following expression with respect to the age at first reproduction (Tm)

$$\mathrm{LTRO} = \sum_{t=\mathrm{Tm}}^{\mathrm{Tmax}} C \cdot W_t^{d} \exp(-Mt)$$

Here the yearly mortality rate (M) is assumed to be constant, W_t refers to the size at age t, and Tmax is the maximum observed age. We examined the sensitivity of the optimal age at maturity and lifetime reproductive output to changes in M, A, and C using the parameter values listed above for Pacific herring and setting M equal to 0.45. The following responses to a 10% increase in the selected parameters were obtained:

Parameter	Tm,opt	LTRO
M	-22%	-24%
A	+10%	+16%
C	~0	+ 2%

The results indicate that both Tm,opt and LTRO are most sensitive to the mortality rate, and suggest that the age (and hence size) at maturity should increase when the mortality rate is low, and decrease when M is high. This is to be expected because M discounts the relative value of energy invested in growth by decreasing the probability of future reproduction. There is empirical support for this trend amongst the species shown in Fig. 9, as body weight, and presumably age, at maturity tend to be negatively correlated with the average natural mortality rate (M). Additionally, the respective reproductive effort coefficients (C/A) of these species tend to be positively correlated with M, indicating that for any given size, species which experience high natural mortality rates have much higher reproductive efforts. Other related correlates with mortality rate (M) have been reported. Empirical relationships between life history parameters (Fig. 10) indicate that the asymptotic length, Lmax, is negatively correlated with M (Adams, 1980; Gunderson, 1980; Pauly, 1980), while the ratio of the length at first maturity to the asymptotic length (Lm/Lmax), indicative of post maturational growth potential and hence reproductive effort, is positively correlated with M. A parallel relationship between M and the ratio Wm/Wmax can be obtained from Fig. 9.

The surplus power coefficient, A, has a greater effect on Tm,opt (and LTRO) than does the reproductive output coefficient, C. This suggests that under a given mortality schedule variations in observed age at first maturity are more likely to be driven by variations in the processes of energy acquisition than energy allocation.

Fig. 10. Reported relationships between various life-history parameters in fishes. Signs indicate signs of correlations. Lmax is the asymptotic length, GSI is the gonosomatic index, Tm is the age at first maturity, Tmax is the maximum observed age, M is the instantaneous annual natural mortality rate, K describes the rate at which an organism grows towards Lmax and completes the growth pattern (von Bertalanffy equation), and Lm is the length at first maturity. Sources: Adams (1980), Beverton & Holt (1959), Gunderson (1980), Pauly (1980), Roff (1984), Ware (1980).

	L_{max}	GSI	T_m	T_{max}	M	K
GSI	−					
T_m	+	−				
T_{max}	+	−	+			
M	−	+	−	−		
K	−	+	−	−	+	
$\frac{L_m}{L_{max}}$	−	+	−	−	+	+

Reproductive Cost

When mortality rate is assumed to be constant or to decrease subsequent to sexual maturity, the curve of lifetime reproductive output with respect to the age at first maturity is flat-topped with a poorly defined optimum (e.g. Fig. 13 in Jones & Johnston, 1977). However, in most fish species there is a tendency for the natural mortality rate to be U-shaped with respect to age. It descends very rapidly during the early life history and then flattens out. Around the age at maturity the mortality rate begins to rise, at first slowly, but then very rapidly. In addition, there is evidence for some species that this rise is size-dependent, that is the apparent mortality cost of reproduction is higher the longer reproduction is delayed (Ware, unpublished data). By using such size-dependent U-shaped mortality functions we obtained a more sharply defined optimum age at maturity, indicating that this characteristic may be more constrained by selection than other analyses suggest (e.g. Jones & Johnston, 1977).

PERSPECTIVE

One of the primary factors leading to the evolution of large pelagic predators has been the development of the high resolution eye of the coleoid cephalopods and vertebrates. This eye resolves better than any compound eye and presumably set the stage for the evolutionary breakthrough in long range detection necessary to support large body size. Land (1984) observed that "for animals separated from ourselves genetically by at least 600 million years of evolution, the similarities in eye design and eye use are impressive enough for one to believe that for a large predator, at least, there is only one way of designing a system that sees well". The threshold body size where this system first appears in cephalopods and fish (1 to 5 mm) is remarkably close to where we calculate it ought to occur (Fig. 5).

By considering the energy gains and costs of visual predators we identified two other ecologically significant sizes: that maximizing the rate of energy gain (EMS) and that minimizing foraging time (TMS). The EMS is not particularly well determined and occurs largely as a consequence of light attenuation, which limits the maximum underwater sighting range. When mortality rate is low this may be the optimum size to terminate growth and allocate all surplus energy to reproduction (Sebens, 1979). In contrast the TMS is the least sensitive of our criteria to variations in the foraging parameters and occurs between 15 and 25 cm (Fig. 8). The fact that this range overlaps the adult size of a large number of pelagic fishes suggests that it may be a very significant point in the size spectrum.

Within the range bounded by the minimum size required to support an eye with high resolving power, and a maximum size limited by the light attenuation properties of the environment, realised adult size (size at maturity to asymptotic size) is determined by the way in which energy is allocated between growth and reproduction. These allocation schedules are themselves a consequence of the interplay between natural selection and the processes of foraging, reproduction, and mortality. Although this complex of energy constraints and reproductive trade-offs limiting body size has been identified with specific reference to pelagic fishes, all organisms are clearly subject to such limits in one form or another. Our general approach is therefore broadly applicable to other studies concerning the determinants of body size. While our technique of calculating surplus power is less general, it can be modified to a certain extent by substitution of the appropriate expressions for foraging gain. Such specific adaptations of the pelagic model could yield insights regarding the life history correlates of body size in ecologically similar groups such as pelagic mammals or bathy-pelagic fishes.

Financial support for this study was provided by a NSERC of Canada Visiting Fellowship to RLD. M. Healey made useful comments on an earlier version of the manuscript.

REFERENCES

Adams, P.B. (1980) Life history patterns in marine fishes and their consequences for fisheries management. Fish. Bull., 78, 1-12.

Bachman, R.A. (1981) A growth model for drift-feeding salmonids: a selective pressure for migration. In: Salmon and Trout Migratory Behaviour Symposium, Ed. E.C. Brannon & E.O. Salo, pp. 128-135. University of Washington, Seattle.

Bell, G. (1984) Measuring the cost of reproduction. I. The correlation structure of the life table of a plankton rotifer. Evolution, 38, 300-313.

Beverton, R.J.H. & Holt, S.J. (1959) A review of the lifespans and mortality rates of fish in nature, and their relation to growth and other physiological characteristics. In: CIBA Foundation Colloquia on Ageing, Vol. 5, Ed. G.E.W. Wolstenholme & M. O'Connor, pp. 142-180. J. & A. Churchill, Ltd., London.

Blaxter, J.H.S. (1980) Vision and the feeding of fishes. In: Fish Behaviour and Its Use in the Capture and Culture of Fishes, ICLARM Conference Proceedings, Vol. 5. Eds. J.E. Bardach, J.J. Magnuson, R.C. May & J.M. Reinhart. 512 pp. International Centre for Living Aquatic Resources Management, Manila, Philippines.

Calder, W.A. (1984) Size, Function and Life History. Harvard University Press, Cambridge, Mass.

Calow, P. (1979) The cost of reproduction - a physiological approach. Biol. Rev., 54, 23-40.

Charlesworth, B. & Leon, J.A. (1976) The relation of reproductive effort to age. Am. Nat., 110, 449-459.

Confer, J.L. & Blades, P.I. (1975) Omnivorous zooplankton and planktivorous fish. Limnol. & Oceanogr., 20, 571-579.

Dill, L.M. & Fraser, A.H.G. (1984) Risk of predation and the feeding behaviour of juvenile coho salmon (Oncorhynchus kisutch). Behav. Ecol. & Sociobiol., 16, 65-71.

Dunbrack, R.L. (1984) The foraging ecology of juvenile coho salmon: body size, diet selection, and intraspecific competition. PhD thesis. 209 pp. Simon Fraser University, Burnaby, British Columbia.

Dunbrack, R.L. & Dill, L.M. (1983) A model of size dependent surface feeding in a stream dwelling salmonid. In: Predators and Prey in Fish, Eds. D.L.G. Noakes, D.G. Lindquist, G.S. Helfman & J.A. Ward, pp. 41-54. Junk, The Hague.

Eggers, D.M. (1977) The nature of prey selection by planktivorous fish. Ecology, 58, 46-59.

Feigenbaum, D. & Reeve, M.R. (1977) Prey detection in the Chaetognatha : response to a vibrating probe and experimental determination of attack distance in large aquaria. Limnol. & Oceanogr., 22, 1052-1058.

Gibson, R.M. (1980) Optimal prey-size selection by three-spined sticklebacks (Gasterosteus aculeatus): a test of the apparent-size hypothesis. Zeit. fur Tierpsychol., 52, 291-307.

Giguēre, L.A. et al. (1982) Predicting encounter rates for zooplankton: a model assuming a cylindrical encounter field. Can. J. of Fish. & Aquat. Sci., 39, 237-242.

Gilbert, P.W. (1963) The visual apparatus of sharks. In: Sharks and Survival. Ed. P.W. Gilbert. pp. 283-326. Heath, Indianapolis.

Gunderson, D.R. (1980) Using r-K selection theory to predict natural mortality. Can. J. of Fish. & Aquat. Sci., 37, 2266-2271.

Hairston, N.G. et al. (1982) Fish vision and the detection of planktonic prey. Science, 218, 1240-1242.

Hardy, A. (1956) The Open Sea: Its Natural History. Part I: The World of Plankton. Collins, London.

Hart, J.L. (1973) Pacific Fishes of Canada. Fisheries Research Board of Canada Bulletin 180. Fisheries Research Board of Canada, Ottawa.

Hemmings, C.C. (1966) Factors influencing the visibility of objects underwater. In: Light as an Ecological Factor: A Symposium of the British Ecological Society, Eds. R. Bainbridge, G.C. Evans & O. Rackam, pp. 359-373. Blackwell Scientific Publications, Oxford.

Hester, F.J. (1968) Visual contrast thresholds of the goldfish (Carassius auratus). Vis. Res., 8, 1315-1336.

Hyatt, K.D. (1979) Feeding Strategy. In: Fish Physiology, Vol. VIII. Ed. W.S. Hoar & D.J. Randall. pp. 71-119. Academic Press, New York.

Jones, R. & Johnston, C. (1977) Growth, reproduction and mortality in gadoid fish species. In: Fisheries Mathematics, Ed. J.H. Steele, pp. 37-62. Academic Press, London.

Kerfoot, W.C. (1977) Implications of Copepod predation. Limnol. & Oceanogr., 22, 1052-1058.

Koehl, M.A.R. & Strickler, J.R. (1981) Copepod feeding currents: Food capture at low Reynolds number. Limnol. & Oceanogr., 26, 1062-1073.

Land, M.F. (1984) Molluscs. In: Photoreception and Vision in Invertebrates. Ed. M.A. Ali. pp. 699-725. Plenum Press, New York.

Milinski, M. & Heller, R. (1978) Influence of a predator on the optimal foraging behaviour of sticklebacks (Gasterosteus aculeatus L.).

Nature, 275, 642-644.
Myers, R.A. & Doyle, R.W. (1983) Predicting natural mortality rates and reproduction-mortality trade-offs from fish life history data. Can. J. of Fish. & Aquat. Sci., 40, 612-620.
Northmore, D. et al. (1978) Vision in fishes : colour and pattern. In: The Behavior of Fish and Other Aquatic Animals, Ed. D.I. Mostofsky, pp. 79-136. Academic Press, New York.
Pauly, D. (1980) On the interrelationships between natural mortality growth parameters and mean environmental temperature in 175 fish stocks. J. du Cons. Int. l'Explor. de la Mer, 39, 175-192.
Platt, R. & Denman, K.L. (1978) The structure of pelagic marine ecosystems. In: Marine Ecosystems and Fisheries Oceanography Symposia, Rapports et Proces- Verbaux des Reunions, Vol. 173, Eds. T.R. Parsons, B.-O. Jansson, A.R. Longhurst & G. Saetersdal, pp. 63-65. Conseil International pour l'Exploration de la Mer, Copenhagen.
Riessen, H.P. et al. (1984) An analysis of the components of Chaoborus predation on zooplankton and the calculation of relative prey vulnerabilities. Ecology, 65, 514-522.
Roff, D.A. (1983) An allocation model of growth and reproduction in fish. Can. J. of Fish. & Aquat. Sci., 40, 1395-1404.
Roff, D.A. (1984) The evolution of life history parameters in teleosts. Can. J. of Fish. & Aquat. Sci., 41, 989-1000.
Rose, M.R. & Charlesworth, B. (1981) Genetics of life history in Drosophila melanogaster. I. Sib analysis of adult females. Genetics, 97, 173-186.
Schaffer, W.M. (1974) Selection for optimal life histories: the effects of age structure. Ecology, 55, 291-303.
Schmidt-Nielson, K. (1984) Scaling: Why is animal size so important? Cambridge University Press, Cambridge.
Schoener, T.W. (1969) Models of optimal size for solitary predators. Am. Nat., 103, 277-313.
Schoener, T.W. (1971) Theory of feeding strategies. Ann. Rev. Ecol. Syst., 2, 369-404.
Sebens, K.P. (1979) The energetics of asexual reproduction and colony formation in benthic marine invertebrates. Am. Zool., 19, 683-697.
Sheldon, R.W. et al. (1972) The size distribution of particles in the ocean. Limnol. & Oceanogr., 17, 327-340.
Sheldon, R.W. et al. (1973) The production of particles in the surface waters of the ocean with particular reference to the Sargasso Sea. Limnol. & Oceanogr., 18, 719-733.
Walls, G.L. (1963) The Vertebrate Eye and Its Adaptive Radiation. Hafner, New York.
Ware, D.M. (1972) Predation by rainbow trout (Salmo gairdneri): the influence of hunger, prey density, and prey size. J. Fish. Res. Bd. Can., 29, 1193-1201.
Ware, D.M. (1973) Risk of epibenthic prey to predation by rainbow trout (Salmo gairdneri). J. Fish. Res. Bd. Can., 30, 787-797.
Ware, D.M. (1978) Bioenergetics of pelagic fish: theoretical change in swimming speed and ration with body size. J. Fish. Res. Bd. Can., 35, 220-228.
Ware, D.M. (1980) Bioenergetics of stock and recruitment. Can. J. Fish. Aquat. Sci., 37, 1012-1024.
Werner, E.E. et al., (1983) An experimental test of the effects of predation risk on habitat use in fish. Ecology, 64, 1540-1548.
Werner, E.E. & Hall, D.J. (1974) Optimal foraging and the size selection of

prey by the bluegill sunfish (Lepomis macrochirus). Ecology, 55, 1042-1052.

Williams, G.C. (1966) Natural selection, the costs of reproduction, and a refinement of Lack's hypothesis. Am. Nat., 100, 687-690.

THE EVOLUTION OF MAMMALIAN ENERGETICS

B.K. McNab

INTRODUCTION

Few detailed examinations of the evolution of physiological function have been made for, unlike morphological traits, physiology appears to leave few telltale traces of its evolutionary history among contemporary organisms (unless they occur in the form of biochemical pathways), and still less direct evidence is left of its evolution in the fossil record (although see the morphological evidence used by various sides in the dinosaur thermal biology controversy; Thomas and Olson, 1980). A persistent interest in the evolution of function generally depends on the observation of contemporary species, the conclusions being applied to the historic record, assuming that the relationships found in contemporary species were generally present throughout the phyletic history of the species. Of course, such an analysis is fraught with many difficulties, the most striking being a superficial analysis of contemporary species and the assumption that the diversity of contemporary patterns includes all those found in extinct species. That such assumptions are partially unjustified is well-illustrated by the controversy on whether dinosaurs were hot- or cold-blooded (see Thomas and Olson, 1980): there are no terrestrial vertebrates today that are known to have the thermal biology of dinosaurs, if only because few vertebrates have the mass of dinosaurs, so all suggestions on their thermal biology are obtained by extrapolation either from high-intensity endotherms or from low-intensity ectotherms. Dinosaurs may not have belonged to either group. Furthermore, all extrapolations over a ten-fold increase in body mass are surely wrong in detail, and often in generality. These limitations to our understanding of the evolution of function suggest caution, if not in speculation, at least in their acceptance. Caution should apply here as well.

In this chapter the evolution of mammalian energetics will be examined in detail. There is a large set of comparative data available on energy expenditure; it is of importance for many other functions and behaviours; and it is directly coupled to the environment. Variation in the level of energy expenditure in living mammals will be examined with respect to various characteristics of mammals and of their environments, then some of the physiological and ecological consequences of this variation will be described, and finally these interactions will be applied to the known evolution of mammals.

FACTORS ASSOCIATED WITH THE LEVEL OF ENERGY EXPENDITURE

Rate of metabolism is highly variable in mammals, for it may be modified by ambient temperature, activity, and reproductive state, as well as by many other factors. A detailed analysis of all of these interactions would require a treatise unto itself, but a convenient, if simple, way around this complexity is to note that for most species the maximal rate of metabolism is correlated with the resting rate of metabolism and activity metabolism, like basal rate, is scaled to body mass (Hemmingsen, 1960; Lechner, 1978; Taylor et al., 1982; Garland, 1983). That is, a limit to the steady-state factorial scope for metabolism exists (Jansky, 1962; McNab, 1980), so that an analysis of the factors correlated with resting rate of metabolism, especially basal rate of metabolism, will account for much of the diversity in mammalian energetics. Most of the variation in basal rate of metabolism is correlated with body mass, food habits, activity level, and with the climate in which mammals live, each of which will be examined briefly in turn.

Body Mass

The one factor that can "account" for most of the variation in basal rate in mammals is body mass. The general form of this power relationship has been described many times (e.g., Kleiber, 1932; Brody & Proctor, 1932; Benedict, 1936; Kleiber, 1947, 1961; MacMillen & Nelson, 1969; Dawson & Hulbert, 1970): total basal rate is proportional to body mass raised to the ca. 0.75 power. This relationship leads to the well-known observation that mass-specific basal rate is proportional to mass raised to the -0.25 power. Unfortunately, most of the early curves principally used domesticated species, which may well be a physiologically-biased sample that is not applicable to wild species.

A recent compilation of the basal rates of 3 monotremes, 46 marsupials, and 272 eutherians (McNab, submitted) permits a re-examination of this relationship (Fig. 1). When all mammals are combined, mass-specific basal rate is proportional to $g^{-0.287}$, which means that total basal rate is proportional to $g^{0.713}$. This relationship is significantly different from the Kleiber curve and mathematically accounts for approximately 78% of the variation in basal rate. Marsupials collectively have lower basal rates than eutherians, except that much of this difference is due to the low basal rate of small marsupials. If small (mass < 150 g) marsupials are not included in the marsupial regression, the difference between eutherians and marsupials disappears. The ambiguous position of marsupials occurs because all small marsupials enter torpor in association with low rates of metabolism. Thus, the scaling of basal rate is sensitive to factors that interact both with rate of metabolism and with body mass. Such factors include the level and precision of temperature regulation, food habits, and activity level. This analysis suggests that scaling functions described for taxonomically-defined groups tend to reflect the ecological and physiological composition of the groups, rather than any intrinsic scaling relationship or any distinctive taxonomic characteristic of the groups.

The clearest way to describe the effect of mass on basal rate is for groups of mammals to be defined on ecological and functional criteria. This has been accomplished for eutherians with specialist diets (see next section). Under these circumstances the mass-specific scaling power either varies from -0.20 to -0.30, with -0.25 the modular power, or from -0.40 to -0.60. The powers that vary from -0.20 to -0.30 operate over a mass range from 9 to 407,000 g, while higher powers only operate at small body masses (from 3 to 321 g). The lower power range corresponds to the Kleiber scaling relationship, the higher range to a relation that has been referred to as the boundary curve (McNab, 1983), because it separates those mammals that are continuous endotherms from those that have an endothermy that (upon occasion) is discontinuous on a daily basis (see later).

Clearly, basal rate is an exceedingly important correlate of body mass in mammals, but much residual variation remains, which raises the question as to whether scaling itself limits a mammal's basal rate. After all, at masses less than 50 g basal rate may vary by a factor of 4:1, and at masses greater than 10 kg the variation may be as much as 10:1! Furthermore, implicit in fitting metabolism-mass curves is the assumption that the relationship is linear when the data are transformed into logarithms. An examination of Fig. 1 raises some doubts about this assumption (also, see McNab, 1983).

Fig. 1. Logarithm of mass-specific basal rate of metabolism in mammals plotted as a function of the logarithm of body mass. Data from Kleiber (1947), McNab (1978b, 1986a), and the literature. The solid line labelled K is derived from Kleiber (1932, 1947, 1961), that labelled B is from McNab (1983) and the dashed line is the mean curve fitted to all of the data (McNab, submitted).

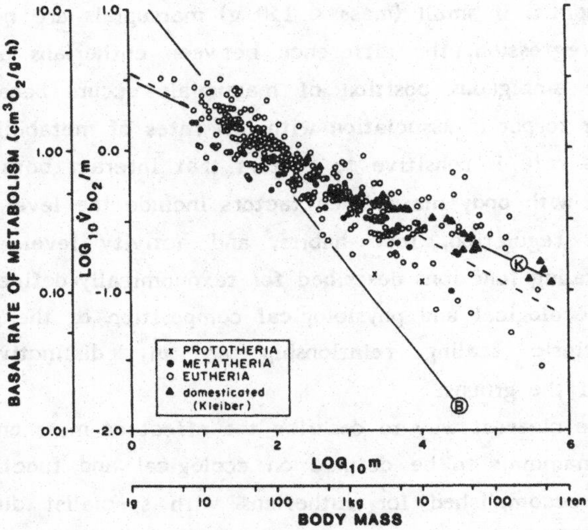

Fig. 2. Logarithm of mass-specific basal rate of metabolism in mammals with specialist diets as a function of the logarithm of body mass (McNab, 1986a). Other symbols as in Fig. 1.

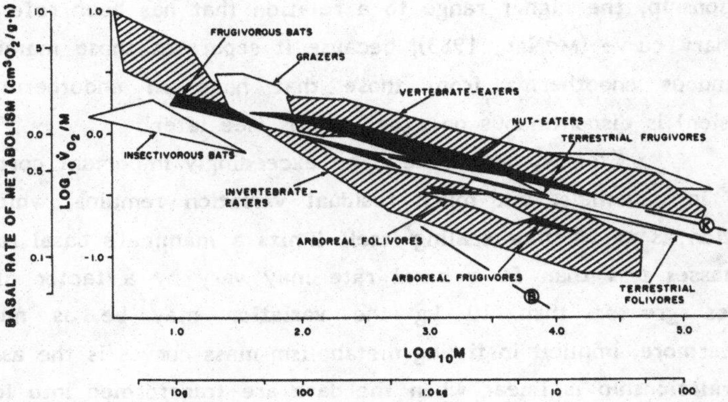

Food Habits

Much of the residual variation in basal rate of metabolism among mammals (after the correlation with body mass is eliminated) is associated with food habits (Fig. 2). The influence of food habits, itself, varies with mass: relatively little variation of basal rate occurs with food habits at small masses, but the correlation with food habits increases with mass, so that most of the variation in basal rate in mammals weighing more than 10 kg is associated with food habits. At intermediate to large masses, mammals that are grazers, vertebrate-eaters, and (probably) nut-eaters have basal rates that equal, or exceed, the Kleiber curve, while those that are browsers and that feed on invertebrates, or on fruit, have low basal rates (McNab, 1986a). The influence of food habits is exaggerated at large masses because total rate of metabolism increases with mass, thereby exaggerating the impact of food availability and quality on the level of energy expenditure. The correlation of basal rate with food habits can be further refined. Thus, not only do intermediate-to-large invertebrate-eaters have low rates, but ant-eating specialists appear to have lower rates than termite-eating specialists (McNab, 1984). Mammals that have mixed diets generally have basal rates that are intermediate to those found in species that are specialised to feed on the components of the mixed diet (McNab, 1986a).

Activity Level

Some mammals are sedentary. They are usually limited to arboreal habits because terrestrial species with such behaviour would be easy prey. The only terrestrial mammals that approach this behaviour have protective structures, such as plates (armadillos, pangolins) and spines (tenrecs, hedgehogs, porcupines, echidnas). Sedentary arboreal species generally live in the tropics where they feed on leaves and/or fruit: these species have lower basal rates than terrestrial species with similar food habits (Fig. 2), apparently as a consequence of small muscle masses (McNab, 1978c).

Climate

Most mammals that live in cold, especially polar, climates have high basal rates. These high rates may directly reflect an adaptation to climate, or they may reflect food habits, because most polar marine mammals are carnivores and most polar terrestrial mammals are grazers or carnivores, food habits that lead to high basal rates. (Baleen whales may have high basal rates, even though they feed on small fish, krill, and other invertebrates, which unlike terrestrial invertebrates, can easily be separated from their matrix, water, and thus may be associated with high basal rates.) As evidence of a direct influence by the thermal environment on basal rate, the monotreme with the highest rate is the platypus, which is aquatic. Heat loss from a warm body in water is high due to the high heat capacity and thermal conductivity of water, so high rates of metabolism in aquatic species may ensure their thermal independence. In fact, the only aquatic mammals known to have very low rates of metabolism are manatees (and by implication, dugongs), the low rates being facilitated by a large mass, and are associated with herbivory; these sirenians are limited in distribution to warm water (Irvine, 1983). Unfortunately, the one recent dugong to live in cold water, the Steller's sea-cow (Hydrodamalis gigas), is extinct. Surely its tolerance to cold water was associated with a mass that reached 4 tons.

Burrowing Activity

The one factor that overrides the correlation of basal rate with food habits is burrowing activity; all burrowing mammals that weigh more than 100 g have low basal rates irrespective of food habits (McNab, 1966, 1979). These low rates are associated with overheating in a burrow system (McNab, 1966; Contreras, 1983) and with living in an atmosphere characterised by low oxygen tensions (Darden, 1972; Withers, 1978).

In conclusion, nearly all variation in the basal rate of living mammals is associated with variations in their body size, food habits, and activity level, and with climatic conditions that mammals face in the environment. When these factors are taken into consideration, the residual variation in basal rate may be as low as 6% in ant/termite-eaters (McNab, 1984), and as low as 1 to 2% in terrestrial browsers and nut-eaters (McNab, 1986a). A variation from 2 to 6% may simply reflect the magnitude of the differences obtained from independent laboratories making measurements on the same species.

CONSEQUENCES OF VARIATION IN RATE OF METABOLISM

A marked variation in rate of metabolism has many consequences, especially in endotherms, which regulate body temperature in part by controlling the rate at which energy is expended. Another function that is correlated with level of energy expenditure is rate of reproduction, not simply because reproduction and growth are energy demanding processes, but also because the ability to commit resources to these processes is positively correlated with variation in basal rate at a fixed body mass. Finally, a correlation seems to exist between level of energy expenditure and the ability of mammals to penetrate cold-temperate and polar environments. All of these interactions have significance for contemporary species and, by implication, for extinct species as well. Each interaction will be examined in turn.

Endothermic Temperature Regulation

Variation in basal rate at large masses influences the level at which temperature regulation occurs, but has little effect on the precision of regulation, unless basal rate is reduced to very low levels. For example, some large invertebrate-eaters, such as Zaglossus, Myrmecophaga, Priodontes, and Orycteropus, have low body temperatures, but retain fairly precise temperature regulation (McNab, 1984). This independence of the precision of regulation from basal rate undoubtedly results from the thermal buffering provided by a large mass. A large reduction in basal rate at small masses, however, has a marked effect both on the level and the precision of temperature regulation (McNab, 1970), and it leads to the use of daily torpor (McNab, 1983).

Fig. 3. Logarithm of mass-specific basal rate of metabolism in mammals as a function of body mass and of the entrance, or not, into daily torpor. Data from McNab (1978b, 1983). Other symbols as in Fig. 1.

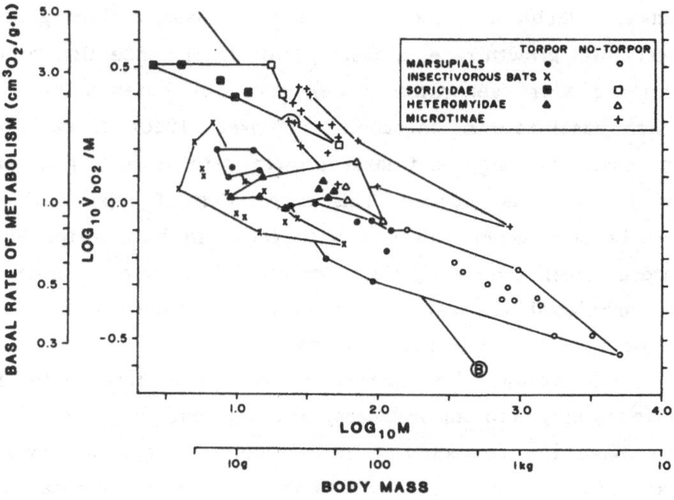

Small endotherms enter daily torpor, unless mass-specific rate of metabolism, in compensation for a small mass, shows an increase beyond that found in the Kleiber curve. These compensatory increases collectively form another scaling relationship, the so-called boundary curve for endothermy (McNab, 1983; labelled "B" in Fig. 3). Mammals can be distinguished by whether they conform to the boundary curve and thereby whether they have very high basal rates at small masses and never enter daily torpor (e.g. most soricine shrews, all microtine rodents), or whether they (at the other extreme) do not conform to the boundary curve and often enter daily torpor (e.g. all small marsupials, all small heteromyid rodents, all insectivorous bats; Fig. 3). Of course, species with an intermediate rate of metabolism occasionally enter daily torpor (e.g. small cricetine rodents, crocidurine shrews, small frugivorous bats). Eutherians obviously have a diverse response to a small mass, apparently reflecting the diverse ecological conditions, including food habits, that eutherians face. This observation is notable in light of the uniformly low basal rate found in small marsupials (Fig. 1), in spite of an appreciable ecological diversity, and of the consequent entrance of all small marsupials into torpor (Fig. 3).

Reproduction

An analysis of the correlation of reproduction with energetics in mammals is complicated by the interactions that occur amongst rate of metabolism, body mass, parameters of reproduction, and the degree of precociality or altriciality of the young (McNab, 1980; Hennemann, 1983, 1984; McNab, 1986b). In eutherians at a given mass, a short gestation period, a high post-natal growth rate, a short period from conception to weaning (and consequently a short generation time), and a high fecundity are associated with a high basal rate of metabolism (McNab, 1980). If basal rate is low, gestation period is long, post-natal growth rate is low, generation time is long, and fecundity is reduced. This correlation of reproduction with basal rate in eutherians extends over a wide range in basal rate. As a result of these interactions, the population exponential growth constant, r_{max}, is positively correlated with rate of metabolism in eutherians independent of the influence of body mass (Hennemann, 1983).

Reflecting the correlation of basal rate with food habits, ant/termite eaters, arboreal folivores, and arboreal frugivores, independent of body size, have low reproductive rates, their young grow slowly, and these species are often characterised by extended periods of parental care, whereas microtine rodents and rabbits have high reproductive rates and short periods of parental care. The species with the greatest fluctuations in population density, thus, are microtine rodents and hares: they have high rates of metabolism and do not enter torpor. These correlations suggest that the principal advantage to a high rate of metabolism in eutherians is a high reproductive rate, which may be further facilitated by evading torpor, although a high rate of metabolism also permits more effective temperature regulation in a cold climate. The comparative importance of a high rate of reproduction and effective temperature regulation are difficult to separate because they are linked together through a high rate of metabolism in eutherians.

In marsupials the correlation between reproduction and rate of metabolism is much weaker than in eutherians. In eutherians it extends from very low rates to rates that are much greater than Kleiber, but in marsupials it extends from low rates to rates that are only about 75% of Kleiber (McNab, 1986b). This limited coupling of reproduction to basal rate in marsupials suggests why no marsupial is known to have a high basal rate of metabolism: marsupials would obtain no increase in reproductive output from a high rate. In other words, the presence of low basal rates in some marsupials and in eutherians appears to be imposed by limitations in the characteristics of food, but the absence of high rates of metabolism in marsupials appears to represent a limitation internal to marsupials (McNab, 1986b).

Distribution

The great malleability of rate of metabolism permits eutherians to move into most environments. Among the most demanding conditions are those found in cold-temperate and polar environments, especially during winter. Eutherians generally tolerate these conditions by having high rates of metabolism, which lead to effective temperature regulation and to an intense period of reproduction produced by large litter sizes, multiple litters, and high growth rates. Reproduction in cold-temperate environments is also enhanced in small species by high rates of metabolism, which reduce entrance into torpor. As has been seen, marsupials are generally not characterised by high growth rates or multiple litters, but are characterised by long periods from conception to weaning (Russell, 1982; McNab, 1986b). These traits may account for the general absence of marsupials from cold-temperate and polar environments, although some marsupials live in seasonally cool-to-cold environments in South America (Caenolestes, Rhyncholestes, Dromiciops,

Lestodelphis), North America (Didelphis), and Tasmania, in part through the use of seasonal torpor (if small) or associated with a large mass.

THE EVOLUTION OF MAMMALIAN ENERGETICS

Mammals were derived, apparently only once, from cynodont therapsids (Hopson, 1969; Hopson & Crompton, 1969; Crompton & Jenkins, 1979). The reptilian Order Therapsida was highly diverse in terms of food habits, morphology, and taxonomy. Most species were medium to large in size, which probably meant that they were to some degree homoiothermic, i.e., they had an approximately constant body temperature provided, primarily, by the thermal inertia of a large mass (Spotila et al., 1973; McNab & Auffenberg, 1976). Thermal stability in therapsids may have been enhanced through the acquisition of a fur coat, by vasomotion, and through the behavioural control of heat exchange. de Ricqles (1974, 1980) showed that therapsids had a bone structure typified by haversian systems, which today is found in homoiotherms, not by the annual pattern of bone deposition that is found in living poikilotherms. These observations led Bakker (1972) and de Ricqles (1974, 1980) to conclude that therapsids were not only homoiothermic (i.e., that they had a constant body temperature), but that this constancy was based on high (mammalian-like) rates of metabolism. Yet, there is no direct evidence of such high rates, and the argument that endothermy is required for thermal constancy is a poor analogy with small mammals, for large mammals, as seen, do not require high basal rates for temperature regulation (Fig. 1). Most therapsids probably were inertial homoiotherms, but these reptiles most likely had low rates of metabolism, probably roughly equivalent to those found in living reptiles.

THE CONVERSION OF INERTIAL HOMOIOTHERMY TO ENDOTHERMY

The most striking morphological change to occur in the evolution of mammals from cynodont therapsids was a huge decrease in body mass (Hopson, 1973; McNab, 1978a; Fig. 4). Considering the interaction between basal rate and body mass, the decrease in body mass was undoubtedly associated with a change in the thermal biology of cynodonts, because small species cannot be inertial homoiotherms. This conclusion gains force from the observation that the decrease in mass associated with the evolution of mammals may have been as great as 1000:1 (McNab, 1978a)! Homoiothermy can be transferred to a small mass, if a compensatory increase in rate of

Fig. 4. Skull length in cotylosaurs, pelycosaurs, therapsids, and mammals as a function of time (measured in millions of years before the present). Modified from McNab (1978a).

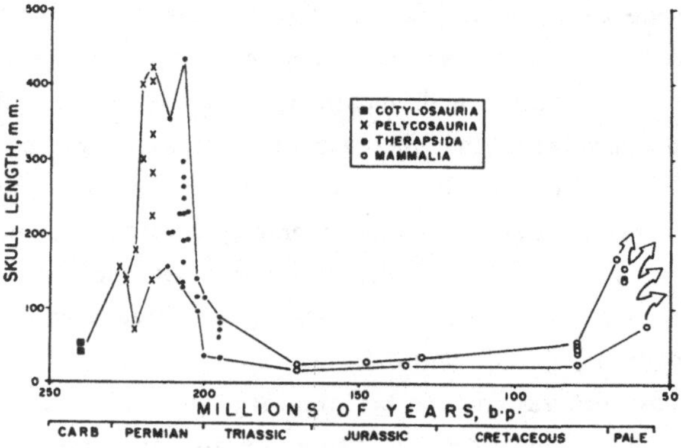

Fig. 5. Logarithm of total rate of metabolism as a function of the logarithm of body mass on eutherians (assuming that they do not enter torpor), marsupials, and reptiles at a body temperature of 20 and 30°C. The minimal boundary curve for endothermy and a possible pathway (shaded) by which the presumptive inertial homoiothermy is converted to the suggested endothermy of early mammals are indicated, as is the suggestion that with the evolution of eutherians an increase in rate of metabolism occurred that led to the elimination of "obligatory" torpor. Data from Bennett & Dawson (1976) and McNab (1983, submitted).

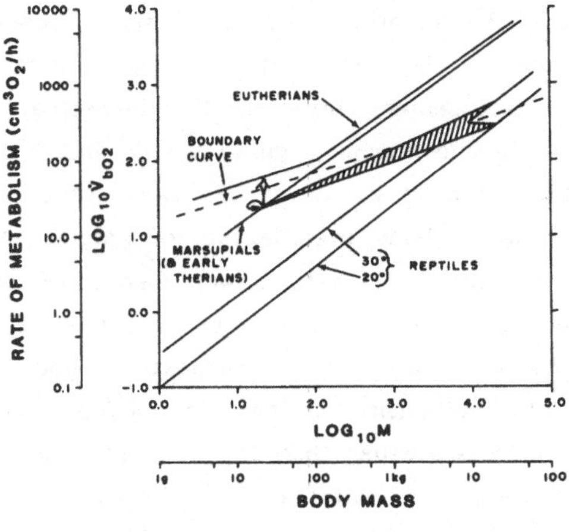

metabolism occurs relative to the standard scaling relations. In other words, small vertebrates cannot be homoiothermic unless they are endothermic.

Cynodonts show anatomic evidence of a compensatory increase in rate of metabolism connected with the reduction in body mass: as mass decreases, a secondary palate progressively developed in the skull of cynodonts (McNab, 1978a). A secondary palate permits gas exchange to be separated from food intake and mastication, a separation that becomes increasingly important with the increase in gas exchange as food intake increases. Ventilation rate increases with a reduction in mass and with a shift from ectothermy to endothermy, although among ectotherms, such as lizards, a decrease in mass alone is insufficient to compel the development of a secondary palate. A palate also permits suckling and breathing to occur at the same time (Lillegraven, pers. comm.).

These observations can be integrated into a picture of the means by which the presumptive inertial homoiothermy of large therapsids was converted gradually into the likely endothermy of late cynodonts and early mammals (McNab, 1978a). This view is illustrated in Fig. 5. Here the mean curves for total basal rate in eutherians (assuming that they do not enter torpor), total basal rate in marsupials, and total standard rate (at 20 and 30°C) in reptiles are plotted as a function of mass. Note that at all masses mammals have higher total rates than reptiles. Assuming that large therapsids conformed to the reptile curve, the derivation of intermediate-sized cynodonts and small mammals would involve a <u>reduction</u> in total rate of metabolism. As long as the reduction in rate of metabolism with a decrease in mass is less than the decrease expected from the reptilian scaling relation, a large ectotherm would approach the endotherm curve as mass decreases. Such an intermediate scaling function is described by the boundary curve for endothermy. Given the displacement between the marsupial and reptilian curves, a cynodont following the boundary curve would shift to the marsupial curve if body mass decreased by a factor of ca. 250:1, which is well within the estimated factor by which body mass actually reduced. Of course, the earliest mammals may have had basal rates lower than those found in living marsupials - they might have been at the level of living tenrecs. Because eutherians had not yet evolved, however, the earliest mammals were unlikely to have had high rates of metabolism. As indicated in Fig. 5, cynodonts may have followed a curve that was steeper than the boundary curve, which would have meant that the propensity to enter daily torpor would have increased with a decrease in mass.

The evolution of endothermy clearly was a complex change forced by the integrated influence of many factors working in concert. The acquisition of endothermy not only permitted thermal constancy to be transferred to a small mass, but it also was associated with the switch from a dependence on anaerobic to aerobic metabolism during activity, and with the conversion of a sedentary to an active life style (Bennett & Ruben, 1979; Pough, 1980). Endothermy furthermore permitted the coupling of an active life style to nocturnal habits (Hopson, 1973).

THE EARLIEST MAMMALS

The earliest mammals can be characterised as being small (generally less than 100 g, often less than 30 g), furred, nocturnal, and feeding their young by lactation, although they probably laid eggs (Hopson, 1973). Two striking observations are that the first two-thirds of mammalian existence was in the Mesozoic (see Lillegraven et al., 1979) and that mammals remained very small for nearly 100 million years (Fig. 4).

The physiological characteristics of the earliest mammals are difficult to define with certainty, but they probably had low basal rates (recall that only eutherians today have high basal rates), possibly similar to those found in some small dasyurids or small tenrecs, i.e. approximately 40 to 60% of the Kleiber value. Such low rates in mammals that weight less than 100 g means that they would have fallen below the boundary curve (Fig. 3); consequently, all early mammals probably entered daily torpor. Regulated body temperatures in these mammals may have been as low as 32 to 34°C, which is typical of the small tenrec *Microgale talazaci* (Eisenberg & Gould, 1970).

The earliest mammals, like their reptilian forebearers, were undoubtedly oviparous, as are living monotremes. The shift to viviparity probably occurred at the end of the Jurassic with the appearance of the first Theria, the direct ancestors of the viviparous metatherians and eutherians, or possibly in the Eupantotheria or Symmetrodonta. Yet, there is no assurance that all Prototheria were egg layers. In fact, Kielan-Jaworowska (1979) argued that the pelvic girdles of multituberculates indicated that they gave birth to very small neonates, possibly similar to living marsupials. Whenever the appearance of viviparity occurred, the strong positive coupling of reproduction to energetics must have waited until the appearance of the Eutheria. Only then were small mammals capable of increasing basal rate sufficiently for them to evade the torpor that until then was an obligatory consequence of a small mass among mammals and their cynodont ancestors (Fig. 5). At that time mammals, in the form of eutherians, were finally able to exploit fully those resources that permitted especially high rates of reproduction and, possibly for the first time, to occupy fully cold-temperate and polar environments.

THE EVOLUTIONARY HISTORY OF MAMMALS AND OF THE VARIATION IN THEIR BASAL RATES

The historic pattern by which rate of metabolism varied in relation to the history of mammalian evolution can now be suggested (Fig. 6). All living monotremes and all marsupials studied have low basal rates by Kleiberian standards. In fact, only eutherians have high basal rates and these rates are found in most living eutherian orders. Those eutherian orders that are characterised by low basal rates have low rates because these eutherians have a body mass greater than 50 g and use invertebrates (Xenarthra, Pholidota, Macroscelidia, Tenrecomorpha, Erinaceomorpha, Scandentia, Tubulidentata), or the leaves of woody plants (Xenarthra, Hyracoidea), as the

Fig. 6. The evolutionary relationships amongst mammals, living and extinct (derived from McKenna, 1975). The following symbols are used: +, extinct; ? before group, doubt about ancestry; ? after group, no measurements of basal rate; * after group, one or more species of this group have basal rates that equal, or exceed the values expected from the Kleiber curve; a is the earliest point and b the latest point at which reproduction could have become strongly coupled to energetics and, thus, the points at which Kleiberian rates of metabolism could have evolved.

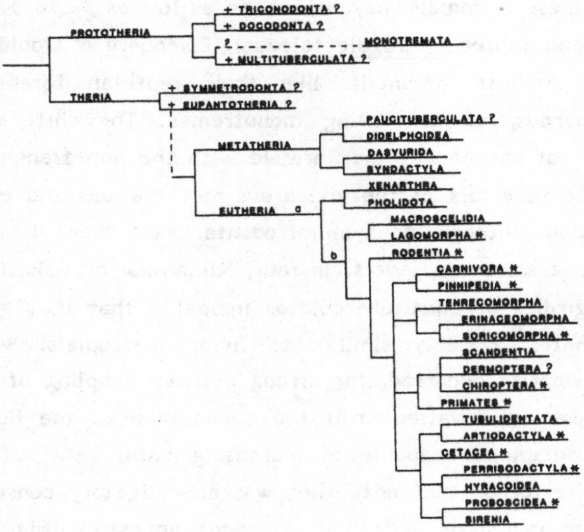

principal foods. Because of the evolutionary relationships suggested in Fig. 6 and the occurrence of high basal rates, the coupling of reproduction to energetics as found in most eutherians probably occurred either at the origin of the Eutheria (point a), or at the latest, after the separation of the "paratherians" (= Xenarthra and Pholidota) from the eutherian line (point b).

SELECTIVE TEMPORAL PERSISTENCE OF MAMMALS

One of the most fascinating aspects of the contemporary mammal fauna is that a rich mixture of mammals exist, some of which represent early groups, often of limited diversity, that have persisted for a long time, and others that represent recent groups that have great diversity. Why do some groups persist for long periods, while others have short life spans? Does persistence reflect the features of a group, or is persistence simply a random event? A close examination of the mammal groups that have persisted suggests that they have habits associated with low basal rates. Thus, noted persistence occurs in (a) invertebrate-eaters (e.g. Xenarthra, Macroscelidia, Tenrecomorpha, Erinaceomorpha), and most especially ant/termite-eaters (Monotremata, Xenarthra, Pholidota, Tubulidentata), (b) fruit-eaters (Didelphoidea, Syndactyla, prosimians, Scandentia), and (c) folivores (Syndactyla, Xenarthra, Dermoptera, prosimians, Hyracoidea). These mammals, in a sense, are protected from "progressive" mammals by using food supplies that require low rates of energy expenditure, thereby denying progressive species the ability to use the positive coupling of reproduction to rate of energy expenditure to displace "conservative" species. Mammals that use foods that permit a high rate of metabolism (e.g. grass and vertebrates) have no such protection, and are readily replaced by new evolutionary experiments that can exploit the coupling of reproduction to energetics more effectively than established lines. The only places where conservative species continue to exploit foods that permit high basal rates are on islands or island-continents, where the conservative lines are protected by geographic isolation (McNab, 1986b). Thus, marsupials are the grazers and vertebrate-eaters on Australia and associated islands, and marsupials were the principal vertebrate-eaters on South America until South America was connected with Central and North America in the late Pliocene.

SUMMARY

Variation in mammalian energetics is examined in the context of mammalian history and as a consequence of the impact of environmental conditions on the habits of mammals. The level of energy expenditure in living mammals depends on the level of activity and the costs of reproduction, temperature regulation, and maintenance at thermoneutral temperatures. The principal factor influencing the level of energy expenditure is body mass. Much of the residual variation in energy expenditure is associated with the climate in which mammals live, the foods that they use, and the particular behaviours that they have. The original acquisition of endothermy in the phylogeny of mammals was associated with the marked evolutionary decrease in cynodont body mass. Most marginal forms of temperature regulation in mammals reflect the correlation of rate of energy expenditure with body mass and food habits, not with a phylogenetically "conservative" state. An exception is the absence in marsupials of an adjustment of energy expenditure relative to a small mass, and the resulting ubiquitous occurrence of torpor at small masses. Endothermy may have been perfected at small masses with the coupling of reproduction to energetics in eutherians. Most morphologically conservative mammals that persist use foods that require low rates of metabolism: these foods deny morphologically "progressive" mammals a competitive advantage over conservative species. The only exceptions in which conservative species with low rates of metabolism persevere in eating foods that would permit high rates of metabolism occur on islands and island-continents, where they are protected from eutherians by geographic isolation.

I greatly appreciate the critical reviews of this chapter by J.A. Lillegraven, S.D. Thompson, S.D. Webb, C.A. Woods, and two anonymous reviewers.

REFERENCES

Bakker, R.T. (1972) Anatomical and ecological evidence of endothermy in dinosaurs. Nature, 238, 81-85.

Benedict, F.G. (1936) Vital Energetics. Publ. Carnegie Institution. 503.

Bennett, A.F. & Dawson, W.R. (1976) Metabolism. In: Biology of the Reptilia, Vol. 5. Eds, C. Gans & W.R. Dawson, pp. 127-223. Academic Press, New York.

Bennett, A.F. & Ruben, J.A. (1979) Endothermy and activity in vertebrates. Science, 206, 649-654.

Brody, S. & Procter, R.C. (1932) Relation between basal metabolism and mature body weight in different species of mammals and birds. Missouri Agr. Station Res. Bull., 166, 89-101.

Contreras, L.C. (1983) Physiological ecology of fossorial mammals: a comparative study. Ph.D. Dissertation, Univ. Florida.

Crompton, A.W. & Jenkins, F.A. Jnr. (1979) Origin of mammals. In: Mesozoic Mammals, Eds, J.A. Lillegraven, Z.Kielan-Jaworowska and W.A. Clemens, pp. 59-73. Univ. Calif. Press, Berkeley.

Darden, T.R. (1972) Respiratory adaptations of a fossorial mammal, the Plains pocket gopher (Thomomys bottae). J.Comp.Physiol., 78, 121-137.

Dawson, T.J. & Hulbert, A.J. (1970) Standard metabolism, body temperature and surface areas of Australian marsupials. Am. J. Physiol., 218, 1233-1238.

de Ricqles, A.J. (1974) Evolution of endothermy: historical evidence. Evol.Theory, 1, 51-80.

de Ricqles, A.J. (1980) Tissue structures of dinosaur bone. In: A Cold Look at the Warm-blooded Dinosaurs, Eds, D.K. Thomas & E.C. Olson, pp. 103-139. AAS Selected Symp., No. 28. Amer. Assoc. Advance Sci., Washington.

Eisenberg, J.F. & Gould, E. (1970) The tenrecs: a study in mammalian behaviour and evolution. Smithsonian Contrib. Zool., 27, 1-138. Smithsonian Inst. Press, Washington D.C.

Garland, T. Jr. (1983) Scaling the ecological cost of transport to body mass in terrestrial mammals. Am.Nat., 121, 571-587.

Hemmingsen, A.M. (1960) Energy metabolism as related to body size and respiratory surfaces, and its evolution. Rept.Steno Mem. Hosp., 10, 1-110.

Hennemann, W.W. III (1983) Relationship among body mass, metabolic rate, and the intrinsic rate of natural increase in mammals. Oecologia, 56, 104-108.

Hennemann, W.W. III (1984) Intrinsic rates of natural increase of altricial and precocial eutherian mammals: The potential price of precociality Oikos, 43, 363-368.

Hopson, J.A. (1969) The origin and adaptive radiation of mammal-like reptiles and nontherian mammals. Ann. N. Y. Acad. Sci., 167, 199-216.

Hopson, J.A. (1973) Endothermy, small size, and the origin of mammalian reproduction. Am.Nat., 107, 446-452.

Hopson, J.A. & Crompton, A.W. (1969) Origin of mammals. In: Evolutionary Biology, Vol. 5, Eds, T. Dobzhansky, M.K. Hecht, and W.C. Steere, pp. 15-72. Appleton-Century-Crofts, New York.

Irvine, A.B. (1983) Manatee metabolism and its influence on distribution in Florida. Biol.Conservation, 25, 315-334.

Jansky, L. (1962) Maximal steady state metabolism and organ thermogenesis in mammals. In: Comparative Physiology of Temperature Regulation, Eds, J.P. Hannon & E. Viereck, pp. 175-195. Arctic Aeromed. Lab., Fort Wainwright, Alaska.

Kielan-Jaworowska, Z. (1979) Pelvic structure and nature of reproduction in Multituberculata. Nature, 277, 402-4403.

Kleiber, M. (1932) Body size and metabolism. Hilgardia, 6, 315-353.

Kleiber, M. (1947) Body size and metabolic rate. Physiol. Rev., 27, 511-541.

Kleiber, M. (1961) The Fire of Life. John Wiley & Sons, New York.

Lechner, A.J. (1978) The scaling of maximal oxygen consumption and pulmonary dimensions in small mammals. Resp.Physiol., 34, 29-44.

Lillegraven, J.A. et al., (1979) Mesozoic Mammals. University of California Press, Berkeley.

MacMillen, R.E. & Nelson, J.E. (1969) Bioenergetics and body size in dasyurid marsupials. Am.J.Physiol., 217, 1246-1251.

McKenna, M.C. (1975) Toward a phylogenetic classification of the Mammalia. In: Phylogeny of the Primates, Eds, W.P. Luckett & F. Szalay, pp. 21-46. Plenum Press, New York.

McNab, B.K. (1966) The metabolism of fossorial rodents: a study of convergence. Ecology, 47, 712-733.

McNab, B.K. (1970) Body weight and the energetics of temperature regulation. J.Exp.Biol., 53, 329-348.

McNab, B.K. (1978a) The evolution of endothermy in the phylogeny of mammals. Am.Nat., 112, 1-21.

McNab, B.K. (1978b) The comparative energetics of neotropical marsupials. J.Comp.Physiol., 125, 115-128.

McNab, B.K. (1978c) Energetics of arboreal folivores: physiological problems and ecological consequences of feeding on an ubiquitous food supply. In: The Ecology of Arboreal Folivores, Ed, G.G. Montgomery, pp. 153-162. Smithsonian Institution Press, Washington, D.C.

McNab, B.K. (1979) The influence of body size on the energetics and distribution of fossorial and burrowing mammals. Ecology, 60, 1010-1021.

McNab, B.K. (1980) Food habits, energetics, and the population biology of mammals. Am.Nat., 116, 106-124.

McNab, B.K. (1983) Energetics, body size, and the limits to endothermy. J.Zool., 199, 1-29.

McNab, B.K. (1984) Physiological convergence amongst ant-eating and termite-eating mammals. J.Zool., 203, 485-510.

McNab, B.K. (1986a) The influence of food habits on the energetics of eutherian mammals. Ecol. Monographs, 56, 1-19.

McNab, B.K. (1986b) Food habits, energetics, and the reproduction of marsupials. J.Zool., 208, 595-614.

McNab, B.K. (submitted) Scaling basal rate of metabolism in mammals.

McNab, B.K. & Auffenberg, W. (1976) The effect of large body size on the temperature regulation of the Komodo dragon, Varanus komodoensis. Comp.Biochem.Physiol., 55A, 345-350.

Pough, F.H. (1980) The advantages of ectothermy for tetrapods. Am.Nat., 115, 92-112.

Russell, E.M. (1982) Patterns of parental care and parental investment in marsupials. Biol. Rev., 57, 423-486.

Spotila, J.R. et al. (1973) A mathematical model for body temperatures of large reptiles: implications for dinosaur biology. Am.Nat., 107, 391-404.

Taylor, C.R. et al. (1982) Energetics and mechanics of terrestrial locomotion. I. Metabolic energy consumption as a function of speed and body size in birds and mammals. J.Exp.Biol., 97, 1-21.

Thomas, R.D.K. & Olson, E.C. (1980) A Cold Look at the Warm-blooded Dinosaurs. AAAS Selected Symp., No. 28. Amer. Assoc. Advance. Sci., Washington D.C.

Withers, P.C. (1978) Models of diffusion-mediated gas exchange in animal burrows. Am.Nat., 112, 1101-1112.

INDEX